MEDIATING CLIMATE CHANGE

Environmental Sociology

Series Editors: Michael Redclift, King's College London, UK and Dr Graham Woodgate, Institute for the Study of the Americas, School of Advanced Study, University of London, UK.

Environmental sociology and environment social science sub-disciplines more generally have been developing steadily over the last 30 years. Touching on environmental social movements and grassroots activism; environmental citizenship and green consumerism; sustainable development; the social construction of environmental risk and environmental harm; the environment in the media; ecotourism; and climate change, environmental sociology is a very broad and inclusive subject. This new series, edited by Michael Redclift and Graham Woodgate, is international, cross-disciplinary and designed to produce the leading edge publications in their field.

Mediating Climate Change

JULIE DOYLE
University of Brighton, UK

ASHGATE

Published by
Ashgate Publishing Limited
Wey Court East
Union Road
Farnham
Surrey, GU9 7PT
England

Ashgate Publishing Company
Suite 420
101 Cherry Street
Burlington
VT 05401-4405
USA

www.ashgate.com

British Library Cataloguing in Publication Data
Doyle, Julie.
Mediating climate change. -- (Environmental sociology)
1. Communication in climatology. 2. Climatic changes--
Information services. 3. Mass media and the environment.
4. Science news.
I. Title II. Series
363.7'3874-dc22

ISBN-13: 9780754676683

Library of Congress Cataloging-in-Publication Data
Doyle, Julie, 1972-
Mediating climate change / by Julie Doyle.
p. cm. -- (Environmental sociology)
Includes bibliographical references and index.
ISBN 978-0-7546-7668-3 (hardback) -- ISBN 978-0-7546-7669-0 (ebook)
1. Climatic changes--Social aspects. 2. Environmental mediation. 3. Visualization. I. Title.

QC903.D75 2011
363.738'74--dc22

2011007760

ISBN 978-0-7546-7668-3 (hbk)
ISBN 978-0-7546-7669-0 (ebk)

MIX
Paper from responsible sources
FSC
www.fsc.org
FSC® C018575

Printed and bound in Great Britain by the MPG Books Group, UK.

Contents

List of Figures *vii*
Acknowledgements *ix*
List of Abbreviations *xi*

Introduction: Making Climate Change Meaningful 1

PART I: HISTORICISING/THEORISING CLIMATE CHANGE

1 Problematising Science and Environment: Conceptualising Nature, Vision and Time in the Mediation of Climate Change 13

2 Visualising Climate Change: Negotiating the Temporalities of Climate Through Imagery 31

3 Coda: Nature, Vision and Time 73

PART II: MEDIATING/ADDRESSING CLIMATE CHANGE

4 An Emerging Climate Movement: Questioning Values of Environment, Justice and Faith 77

5 Mediating Copenhagen: Communicating Scientific (Un)certainty and the Political (Un)urgency of Climate Action 101

6 Sustainable Consumption?: Reframing Meat and Dairy Consumption in the Politics of Climate Change 123

7 Imaginative Engagements: Critical Reflections on Visual Arts and Climate Change 145

Epilogue: Positive Action in a Changing Climate 157

References *159*
Index *177*

List of Figures

2.1	IPCC graph showing global mean combined land-air and sea-surface temperatures from 1860-1990	35
2.2	Computer modelling showing effects of doubling CO2 for July-August-September on surface air temperatures (10 year means) from the IPCC	36
2.3	Front cover of Greenpeace's *Emerging Impacts of Climate Change?* report	42
2.4	Photograph from inside Greenpeace's *Emerging Impacts of Climate Change?*	44
2.5	Front cover of Greenpeace's *Putting the Lid on Fossil Fuels*	48
2.6	Front cover of Greenpeace's *Nature's Bottom Line: Climate Protection and the Carbon Logic* showing crack in Larsen B ice shelf	52
2.7	IPCC graphs showing visible correlations between observed and simulated temperature rises from 1860-2000	56
2.8	An individual and collective appeal to climate action from Greenpeace	65
3.1	Larsen B ice shelf, Antarctica. First satellite image of the ice shelf taken by NASA on 31 January 2002; second image taken on 5 March 2002	73
4.1	Generic CAFOD poster for advertising local climate events	89
4.2	Climate Camp penknife design	97
6.1	The Food Chain Campaign from Friends of the Earth UK	139

Acknowledgements

Many people have helped me on my journey in writing this book. I'd like to thank the following friends and colleagues for offering their time in reading through various chapters, for their thoughts and their encouragements: Caroline Bassett, Jody Boehnert, Tara Brabazon, Rebecca Bramall, Jenny Elliott, David Harradine, Irmi Karl and Kate O'Riordan. My thanks also go to Alison Anderson, Max Boykoff, Anabela Carvalho, Stephen Depoe and Anders Hansen for a supportive and vibrant environmental communication culture.

The inaugural cohort of students (2009-10) on the undergraduate Environment and Media Studies degree programme (University of Brighton) have been particularly inspiring. Special thanks to Julie Bevis for providing the idea for one of the case studies in the book. Thanks also to the undergraduate Media Studies students (University of Brighton), particularly those on the 'Visual Culture' and 'Science, Technology and Media' modules, and postgraduate students on the 'Mediating the Environment' module, for thoughtful and perceptive commentaries on environmental and climate change communication. Special thanks to former Media student, Elizabeth Batchelor, for also providing inspiration for one of the case studies in the book. Thanks to those who have generously provided me with images for the book, particularly Greenpeace International, Greenpeace UK, Friends of the Earth UK, CAFOD and Ultimate Holding Company.

My appreciation to Michael Balmond and Juliet Williams for sharing penguin jokes, to Emma Gibson and Debs Kocziban for friendship and Greenpeace memories, to Fraser Macdonald for moments of joy, to my Liverpool and German families for looking after me whilst writing during visits and, above all, to Irmi for her tireless support and encouragement throughout the writing process and beyond.

List of Abbreviations

BBC	British Broadcasting Corporation
CAFOD	Catholic Agency for Overseas Development
CAN	Climate Action Network
CCA	Camp for Climate Action
COP15	Copenhagen Conference – Fifteenth Conference of the Parties under the UNFCCC
FAO	Food and Agriculture Organization of the United Nations
FoE UK	Friends of the Earth UK
FoEI	Friends of the Earth International
GCC	Global Climate Coalition
IIED	International Institute for Environment and Development
IPCC	Intergovernmental Panel on Climate Change
IPPR	Institute for Public Policy Research
MFM	Meat Free Monday
NASA	National Aeronautics and Space Administration
nef	new economics foundation
NGO	Non-governmental organisation
OU	Open University
PETA	People for the Ethical Treatment of Animals
Peta2	Youth Division of People for the Ethical Treatment of Animals,
UEA	University of East Anglia
UN	United Nations
UNEP	United Nations Environment Programme
UNFCC	United Nations Framework Convention on Climate Change
WCED	World Commission on Environment and Development
WMO	World Meteorological Organisation

For Mum and Dad, Ingrid and Hans,
and Irmi, with love.

Introduction
Making Climate Change Meaningful

It is an overcast Saturday afternoon in Brighton, UK. I am standing on the street pavement trying to catch the attention of the passing public. The year is 1997. As co-ordinator of the local Brighton and Hove Greenpeace Group, I and other members of the group have set up a campaign stall on the busy streets of Brighton. Our mission is a difficult one; to explain to the public the dangers of global warming, through Greenpeace UK's campaign to stop new oil exploration in the Atlantic Ocean, off the northwest coast of Scotland. The easy part is explaining to the public that this area of the ocean and the islands of St Kilda are home to a variety of seabirds, seals, whales, dolphins and cold-water coral that are unique to this habitat. Oil exploration in this vicinity would undoubtedly harm these animal and plant species. The beautiful photographic images of gannets, dolphins and seals that we show to the public convey the pristine visual habitat and help garner their support. But the difficult aspect of our campaigning efforts is linking oil exploration, and the burning of fossil fuels, to the production of CO2 and global warming. Most people do not seem to understand global warming – its causes or effects – nor believe it is a reality. The recurrent response from the public is that they cannot see it, so therefore it cannot be happening.

Much has changed since 1997; the year the Kyoto Protocol was established. Since then, the increasingly visible impacts of climate change (a term now more commonly used than global warming) (Whitmarsh 2009) have been circulated by environmental NGOs and the news media, through iconic images of the world's melting glaciers, retreating polar ice caps, 'stranded' polar bears, floods, droughts and hurricanes (Doyle 2007, Manzo 2010). Films such as Al Gore's *An Inconvenient Truth* (2006), along with the rise in media coverage of climate change since the mid 1990s (Boykoff and Roberts 2007), have undoubtedly contributed to the growing public and political awareness. If the public may not fully understand the complexities of climate change, the majority have accepted it as a reality (Leiserowitz et al. 2008; World Development Report 2009). And while the 'Climategate' controversy involving the leaked emails from the Climatic Research Unit at the University of East Anglia (UK) in November 2009 led to a decrease in public confidence in the reality of human induced climate change (Spence et al. 2010; Leiserowitz et. al. 2010), one can be hopeful that this is merely a short term reaction. Indeed, what was once a concern primarily for scientists and environmentalists has now become an issue for the world's governments, as well as businesses, aid agencies, development organisations, religious communities and citizens across the globe. Following years of scientific research and persistent political pressure from the environmental movement, climate change has now entered mainstream politics and culture.

And yet, in the midst of this transformation of climate change from being a concern primarily for science and environmentalists, to a social, cultural, economic and political issue, global greenhouse gas emissions continue to rise. As the UK Met Office has shown, 'All the years from 2000 to 2008 have been in the top 14 warmest years on record' (Met Office 2009). The COP15 conference in Copenhagen in December 2009 failed to deliver a legally binding agreement to commit the world's nations to reducing greenhouse gas emissions by 2020 and beyond. Developing countries and civil society groups argue that the historical responsibility for climate change lies with developed countries; as developing countries will be the most immediately affected by global climatic changes, while contributing the least to greenhouse gas emissions. Thus, while international public and political acknowledgement of climate change may be higher than ten years ago, this has not translated into concerted action on a personal, national or global level. While the original gap between scientific consensus on the anthropogenic causes of climate change and public and political acceptance of this issue could be characterised by one of time and perception, the real gap is now increasingly related to the profound disconnect between political and public acceptance of climate change and the transformation of this knowledge into effective, urgent action to both mitigate and adapt.

To understand the urgencies of climate change is not simply a question of understanding or communicating the science better (Hulme 2009). In this book I assert that we need also to understand how climate change is *made meaningful* in order to be able to better address this issue. How climate change is perceived – individually and collectively – depends upon how it is made socially and culturally meaningful to particular audiences. As Maarten Hajer has argued, 'the dynamics of environmental politics cannot be understood without taking apart the discursive practices that guide our perception of reality' (Hajer 1995, 17). Mike Hulme has also argued that, 'the idea of climate change carries quite different meanings and seems to imply quite different courses of action' (Hulme 2009, xxvi). Climate change exists as a physical reality *and* as a range of social and cultural meanings, which guide our perceptions of the issue and the nature and urgency of our (in) actions. As such, knowledge and understanding of climate change is *mediated* through a range of social and discursive practices. The unique complexity of climate change means that such mediations span a wide range of institutions and representational practices, from positivist science and the environmental movement, to media representations, politics and popular culture. All mediations are performative (Chouliaraki 2008), because knowledge and understanding are actively constituted through communicative processes, simultaneously shaping how we are to act on such knowledge.

Mediating Climate Change examines a range of social and political actors involved in the communication and making meaningful of climate change, including scientists, civil society groups, mass media and artists. It argues that we need to understand how knowledge of climate change has been shaped historically by scientists, environmentalists and the mass media in order to better address the current gaps between our knowledge and our actions. A central argument of this book is that one of the difficulties in engaging people with climate change is due to its historical framing as an environmental issue, which has led to a separation of humans and culture from the environment. This separation has been supported by the scientific knowledge systems and representational practices called upon to identify, validate and communicate climate change, which have unwittingly rendered it as a distant and future event rather than a present and immediate reality. I argue that the interrelated concepts of *nature*, *vision* and *time*, and their centrality to the production and communication of scientific and environmental knowledge, need to be interrogated in order to understand the difficulties these have posed in making climate change both intelligible and meaningful. In doing so, the book explores how these historical and conceptual difficulties involved in the production and communication of climate knowledge are being reproduced and/or transformed by contemporary mediated communication on climate change, across a variety of social and political actors.

Conceptualising climate change – identifying the problems

Environmental communication, social science and media studies scholars have shown that how we come to understand and define the environment shapes how we act upon it (Cox 2006). Climate change emerged as an environmental concern during the mid 1980s, yet it is an issue that not only crosses, but calls into question, the conceptual boundaries between humans and environment, nature and culture, because climate change is also a humanitarian issue. Framing climate change in terms of the environment has led to particular ways of understanding and acting upon this issue which have proven problematic for its effective communication, and for mobilising action. The problems concern how we come to understand the environment, and the forms of knowledge called upon to define climate change as an environmental issue. These problems of definition, communication and action are related to the conceptions of nature, vision and time as they emerge within and across the disciplinary and institutional boundaries of environmental and scientific knowledge.

Since the enlightenment, the environment has been conceptualised predominantly as the natural world, situated as a set of objects externalised and spatially separate from humans and culture (Cox 2006). Constitutive of scientific discourse, such nature/culture and human/animal distinctions have been criticised for reinforcing power relations which treat the environment as a commodified and gendered property separate from humans (Adam 1998a; Plumwood 2002). Problematically, in their defence of the environment, environmentalists have reinforced the conceptual division between humans and nature, presenting humans as doing damage to a nature comprised of landscape and animals (Macnaghten and Urry 1998). This problem is evident in the campaigning example I began this introduction with, which illustrates the specific problems such distinctions pose for communication and action on climate change. While the public could comprehend the potential damage of new oil exploration to animal habitats in a specific geographical area, they could not make the associative link between oil and CO2 production that would affect all living beings and ecological systems as a result of climate change.

Central to the conceptual division between humans and nature, as a condition of scientific empiricism (Plumwood 2002; Adam 1998a) and environmentalism (Latour 2004; Shellenberger and Nordhaus 2007), is an investment in the visible as evidence of truth. In my own experience of climate change campaigning, this manifested in two ways. Firstly, the environment is perceived as that which can be seen – animals and landscape – reinforced through photographic images of a nature at risk. Secondly, the public's lack of belief in the reality of climate change was due, in part, to its relative invisibility; they could not see it. Climate change is characteristic of environmental risks within a risk society; its existence and effects are incalculable, unquantifiable, and often invisible (Beck 1992; Allan et al. 2000). The very fact that the politicians and the public largely accept the reality of climate change is, in part, related to the fact that its effects are now increasingly visible and felt, particularly at a local level. Yet the current visibility of the 'symptoms' (Adam 1998a) of climate change reveals a further problem in communicating this issue, that is, the concept of industrial time (Adam 1994, 1998a), related to the notion of visible truth.

Climate change necessitates action before its effects can be seen. Investment in the visible present as a condition of scientific knowledge, reinforced through environmental (and media) discourses which often present a visible nature at risk (Hansen 2000, Cottle 2000, Doyle 2007), has proven problematic for climate change; an issue that is not always visible, whose causes and effects do not follow a linear trajectory, nor are neatly bound in time or space. Barbara Adam has already identified how the temporalities of environmental time are at odds with industrial clock time (1998a, 1998b). Therefore, the scientific and environmental knowledge systems and values called upon to make climate change intelligible are limited by their conceptual foundations. It is a central tenet of this book that in order to understand how best to address the issue of climate change we must recognise the institutionalised knowledge systems and discursive practices through which climate change has come to be identified and made meaningful. If climate change has truly moved beyond the concerns of scientists and environmentalists to become a mainstream issue, but this concern is not translating into positive concerted action, then it is time for our perceptual framework to be questioned and transformed. But change in thought can only come from an understanding of how our thinking about climate change is shaped in the first place.

Purpose of the book

Given the urgent need to address climate change, what is the purpose of an academic book on the mediation of climate change? What effect will it have on mitigating climate change? Why not just use the time to lobby governments into action, do public campaigning to spread the message or perform direct actions? The reason I pose this question is because, as an academic and environmentalist, I ask myself this all the time. Indeed, in posing this question, I am also remembering and reflecting upon my own journey from being involved in public campaigning and direct action on the environment and climate change, to writing as an academic about the discursive and representational practices that shape our perception of climate change. As a local volunteer for Greenpeace from 1994-2002, my initial engagement with climate change was from an activist perspective. Although I was also an academic, I did not want to analyse my environmental activities, I simply wanted to do them, without question. I was happy to deliver the messages set by Greenpeace UK to the local public. I felt like I was making a real change by speaking to the public and being involved in direct actions. This included putting solar panels on BP's headquarters in Aberdeen, Scotland, in 1997, and standing on the forecourts of Esso petrol stations in Brighton during 2001, urging car drivers to boycott ExxonMobil because of its funding of George Bush, branded a 'climate villain' for denying climate change and failing to ratify the Kyoto Protocol (Vidal 2005; Monbiot 2007). Yet, the question about the difficulties and frustrations of communicating climate change to the public kept returning to me.

Once I had stopped co-ordinating the local Greenpeace group, I found myself wanting to reconsider the discourses and practices of climate change communication, particularly the role of the visual as a communication and campaigning tool. This led me to retheorise the reliance of the environmental movement (and Greenpeace in particular) upon the image as a form of visible protest (DeLuca 1999; Lester 2007), which subscribes to the spectacular nature of the news media's coverage of environmental issues (Anderson 1997; Hansen 2000; Cottle 2009), but also reinforces how seeing and visually representing the environment becomes metonymic for understanding and valuing it (Macnaghten and Urry 1998). The unseen, invisible characteristics of climate change (Beck 1992; Adam 1998a, 1998b) have made it difficult for environmentalists to communicate and validate this issue as real, given their reliance upon visual representations as a form of communicative action (Doyle 2007). I may never have thought about this, however, had I not campaigned for Greenpeace, nor had an academic interest in the role of the visual in the construction and communication of scientific knowledge of the body (Doyle 2001).

Drawing on my experiences of environmental campaigning for Greenpeace, and as an academic examining the mediating discourses and practices of art and science in conceptions of embodiment, I found myself re-engaging with climate change communication from a similar theoretical perspective. In doing so, I was able to reflect upon how communicating about climate change is never simply about telling the facts – that as a result of human activities, CO_2 emissions are increasing and in doing so are raising global temperatures – but is about how it is made meaningful and relevant to multiple audiences and their everyday lives. How climate change is given meaning also affects the kinds of actions we undertake, and our perception of the efficacy of those actions. Furthermore, the conceptual tools called upon to communicate climate change – particularly dramatic news images of environmental destruction (Cottle 2009) – are part of the problem. Barbara Adam's recognition 'that theory is political in the sense that it is the basis for value-laden action' (Adam 1988a, 6) is particularly compelling here. Media, communication and cultural studies scholars have a role to play in thinking about how climate change is shaped and given meaning through language, discourse, institutions, media technologies and representational practices, because these mediations affect the kinds of actions we can imagine and are therefore willing to undertake.

It is no secret that those countries that produce the least CO_2 emissions and those who are amongst the world's poorest will be those most affected by climate change and the least able to deal with it. The world as we know it will be fundamentally altered by climate change, with increasingly severe and unpredictable weather patterns and events, conflicts over the use of natural resources, the collapse of infrastructures, and the global migration of people as environmental refugees (IPCC 2007a, 2007b; High Representative and European Commission 2008). While this is a reality, how we understand this reality and choose to act upon this knowledge is shaped by how we perceive the issue. If climate change is framed as an issue of social justice, then the human costs of climate change as a result of the unequal distribution of, and access, to, natural resources, are brought to the fore. Framing climate change as a humanitarian and social justice issue constitutes a moral imperative to act. However, if climate change is framed through a neoliberal, consumerist discourse, where changes to consumer habits and individual lifestyles are presented as the means of mitigating climate change, then arguably, existing individual habits and practices are not fundamentally altered (Hobson 2006), and the true cost of consumption in relation to CO_2 emissions, as well as global power relations, is left ignored (WorldWatch Institute 2010). Discursive practices shape how we perceive climate change and what kinds of actions are promoted and undertaken.

Structure of the book

In order to examine how climate change is made meaningful, this book brings together existing work on the environment and climate change, from the perspectives of media and cultural studies, environmental communication and social sciences. Although I am a media studies scholar, this book is not simply about media coverage of climate change. Instead, this book looks at mediation more broadly to interrogate the intersecting knowledges, discourses and social practices, across science, environment, media, politics and culture. With case studies drawn mainly from the UK, Europe and the USA, I am both presenting as well as critiquing a north western perspective.

The book is structured into two sections. The first section explores the problems that climate change has posed for knowledge and communication, focusing upon the historical interconnections between the discourses, knowledges and practices of science, environmentalism and the media in the definition and communication of climate change. Chapter 1 establishes the theoretical framework by examining the ways in which scientific knowledge has constructed concepts of nature, vision and time which have impeded our ability to understand climate change as a temporally complex, invisible and incremental phenomena. In exploring how these constructions have been reinforced through the representational practices of environmentalism and mass media, it calls for a reconception of the environment as not separate from humans, and for more reflexive forms of knowledge production and communication which acknowledge the uncertainties underpinning climate science and climatic change as a basis for ethical action in the present. Given the centrality of the visual to western media cultures, Chapter 2 offers an historical analysis of the role of images in climate communication from 1990-2007 within and across science, environmentalism and the media. Through an analysis of the IPCC reports, Greenpeace climate campaign materials and BBC television documentaries from this time period, it explores how the temporalities of climate change have been negotiated visually by these actors, to make climate change culturally meaningful in the present, or to defer action by figuring it as a future concern. Chapter 3 provides a short summary of the difficulties climate change has posed for western knowledge systems and representational practices, discussed in the first section, in order to move on to explore of how these are being addressed in contemporary communication on climate change in the second section of the book.

The second section examines a selection of contemporary mediations on climate change. Chapter 4 examines an emerging climate movement through an analysis of the climate campaigns of CAFOD, Friends of the Earth International and Camp for Climate Action. Attentive to the social values and organisational structures of each organisation/network, it considers how climate change is being made meaningful through an appeal to values of faith and justice, transforming it beyond the traditional concerns of environmentalism to engage a more diverse range of people. In the context of the 'Climategate' controversy, Chapter 5 critically analyses UK print media coverage of the Copenhagen Conference in December 2009 to explore how notions of scientific un/certainty are used to promote or undermine climate action. Chapter 6 focuses upon the role of consumption in climate change mitigation, through an analysis of civil society campaigns that target meat or dairy consumption as a means of addressing climate change. Chapter 7 considers the possibilities visual arts offers for new ways of perceiving climate change, through an appeal to our emotions, imaginations and our senses, helping us to envisage new social realities. The epilogue re-establishes the need to break down the distinctions between nature and culture, already evident in some of the contemporary mediations examined, in order to present the environment and climate change not as an entity to be observed, but rather as an integral and prescient part of human culture.

Meaning matters

The reason why climate change has failed to engage the public in any meaningful way is not simply due to the complexity of the issue, although this is certainly an important factor. This book argues that how climate change is made relevant to people's everyday lives, how it engages people through their existing social values and norms, *matters*. Climate change needs to be understood as a concern for the 'here and now', rather than a distant future, 'out there' somewhere. This involves making climate change temporally, spatially and socially meaningful and relevant.

The mainstream environmental movement has already begun to call for a transformation of its own approaches to climate change communication and action (WWF-UK 2008a; Shellenberger and Nordhaus 2007; Hale 2008), through its acknowledgement that climate change can no longer simply be framed as an environmental issue. Climate change is also a humanitarian issue, affecting not only access to natural resources, and impacting upon human health (Potter 2008), but also transforming the socio-cultural practices which form the basis of human societies, changing how live, act, think and feel. As the environmental movement questions its own categorisation of the environment and climate change as one of nature, rather than the complex interrelationship between humans and environment, the kinds of actions that are advocated to address climate change also begin to shift, as a broader coalition of groups, from social justice to faith groups, are called upon to tackle climate change (WWF-UK 2008a; Stop Climate Chaos 2010).

Climate change poses fundamental questions about what it is to be a human within a rapidly changing climate and world, and about what we value, what our norms are, and how these are culturally and socially shaped. As individualism has become the dominant ideology within contemporary neoliberal societies (Beck 1992; Bauman 2001), shaping the symbolic and social structures of institutions and everyday lives, the need to think and act collectively on climate change appears particularly difficult – but not impossible. The collapse of the global economy has already revealed the precarious and unjust system of late consumer capitalism, hopefully generating a more sustained and active critique of the ideologies and practices of consumer capitalism, leading to a more equitable, sustainable and less damaging economic and social system for humans and environment. Yet, the economic crisis dominates the news media, as the more devastating crisis of climate change remains a lesser priority for the world's governments and publics. The climate 'crisis' needs to take centre stage. It is real and present.

If we are to avoid becoming 'voyeurs' of our 'own impending peril', through the 'news spectacle' of increasing climate catastrophe (Cottle 2009, 88), then we need to feel engaged in the present, and embodied as active citizens. Getting the balance right between communicating the dangers of climate change and engaging the public and politicians to act is a difficult one (Moser and Dilling 2004; Risbey 2007). How climate change is defined and how it is given meaning shapes how individuals and groups are called upon to take action. In examining the mediating processes which shape our perceptions of climate change, this book hopes to contribute to their transformation.

PART I
Historicising/Theorising Climate Change

Chapter 1

Problematising Science and Environment: Conceptualising Nature, Vision and Time in the Mediation of Climate Change

> ...although the size of the observed warming is broadly consistent with the predictions of climate models, it is also of similar magnitude to natural climate variability. An *unequivocal* statement that anthropogenic climate change had been detected could *not* therefore be made at the time – Reference to the IPCC's 1990 assessment report (Houghton 2004, 104).

> Warming of the climate system is *unequivocal*, as is now evident from observations of increases in global average air and ocean temperatures, widespread melting of snow and ice and rising global average sea level – IPCC 2007 assessment report (IPCC 2007, 30, my emphasis).

As illustrated by the quotations above, 17 years marks the time span between the publication of the IPCC's first assessment report in 1990 and its fourth assessment report in 2007. This period could be characterised as the development of international scientific consensus on the anthropogenic causes of climate change, moving from a position of relative uncertainty to one of certainty (Oreskes 2004). Set up in 1988 by the WMO (World Meteorological Organisation) and UNEP (United Nations Environment Programme), to 'provide independent scientific advice on the complex and important issue of climate change' (IPCC 2004, Foreword), the IPCC was able to state in its second assessment report in 1995 that 'human activities, including the burning of fossil fuels, land-use change and agriculture, *are* increasing the atmospheric concentrations of greenhouse gases (which tend to warm the atmospheric atmosphere', although 'there are many *uncertainties* and many factors currently limit our ability to project and detect future climate change' (IPCC 1995, 3, my emphasis). By 1995 then, uncertainties lay not in the knowledge that human activity was affecting climate, but rather in relation to the nature, magnitude and specific impacts of future climatic changes.

The notion of scientific certainty/uncertainty has been particularly pertinent within the global politics of climate change; brought to the fore most recently by the 'Climategate' controversy surrounding leaked private emails from the Climatic Research Unit at the University of East Anglia (UEA) in November 2009. Although the emails between climate scientists appeared to suggest that certain climate data was being manipulated to hide a decline in global temperatures which could discredit the consensus on climate change, the scientists were actually discussing temperature data obtained from tree rings, which diverged from the sharp increases recorded in air temperature in the last 150 years (Schiermeier 2010, 286). While the data and email discussions did not challenge the scientific consensus on climate change (Nature 2010; Schiermeier 2010; Science and Technology Committee 2010), and 'were not part of a systematic attempt to mislead' (Science and Technology Committee 2010), the event illustrates the political context in which climate science research is undertaken and the difficulties of communicating scientific uncertainties to policy makers (Shackley and Wynne 1996) and publics. While uncertainty is central to climate science in relation to predicting the extent and regional variations/impacts of future warming (Shackley and Wynne 1996; IPCC 2007; Budescu et al. 2009; Schiermeier 2010), such uncertainty is difficult to communicate because of the political context in which climate science has historically developed.

Following the establishment of the IPCC in 1988, the Global Climate Coalition (GCC) was set up a year later, comprised of powerful USA industries, which included Exxon, Ford and Chevron (Gelbspan 1997). Until 2002, the GCC lobbied the USA Government against reducing greenhouse gas emissions and sought to undermine the credibility of climate science through PR that exaggerated the uncertainties within climate science (Greenpeace UK 2001). The notion of scientific uncertainty was deployed as a political tool to aggressively undermine scientific credibility and to forward the economic interests of global corporations. It worked. Supported by a corresponding lack of coverage of climate change in the news media (Andreadis and Smith 2007), or news reports that misrepresented the scientific consensus on climate change by giving equal voice to climate sceptics as to climate scientists, thereby creating informational bias (Boykoff and Boykoff 2004), the lack of public understanding and political will to address climate change continued, remaining so until only fairly recently. The relatively long time it has taken for the world's media, governments and publics to accept human induced climate change as a reality, is thus historically, politically and institutionally contingent, leading to a concern by climate scientists that discussing uncertainties within climate science will be 'misconstrued' as uncertainty about the existence of human induced climate change (Nature 2010, 269).

The history of the developing international scientific consensus on climate change, and the eventual acceptance of this science by the public and national governments (though to varying degrees), is therefore not simply a linear narrative about scientific uncertainties becoming more certain, as the work of the GCC, the world's media, and the UEA leaked emails demonstrates. Instead, the history of climate science reveals the socio-political nature of scientific knowledge and its communication, where such 'knowledge' is open to ideological contestation and (re)definition from a variety of social, political and economic actors (Beck 1992). In the politics of climate change, these actors have historically taken the form of ongoing interactions between the IPCC, the mass media, fossil fuel lobbies and environmental pressure groups (Newell 2000). As science is called upon to validate the existence of climate change, the status of scientific knowledge as a form of 'truth' is also called into question. Here, the predictive modelling and forecasting upon which climate science is predicated, and upon which many of the uncertainties rest (Shackley and Wynne 1996), is at odds with the logic and methods of classical science as a materialist and objective praxis (Adam 1998a, 26). The unpredictable, irreversible, and not always visible phenomena (Beck 1992; Adam 1994, 1998a, 1998b) that characterises climate change problematises the scientific knowledge system upon which it has been reliant for its detection, validation, and ongoing assessment. The uncertainties of climate change in relation to its magnitude, its explicit impacts, and the rapidity of climatic change, are therefore in opposition to the assumed certainties of science.

Characteristic of environmental hazards in contemporary risk societies (Beck 1992), climate change raises questions not only about the nature of science as a knowledge system and as a set of institutional and methodological practices, but also how the environment is constituted and understood through the discourses and practices of science. At the same time, the ways in which these discourses get reproduced through the mediating practices of other social actors, such as the environmental movement and the mass media, also require understanding. This chapter thus examines the privileged role of science as a set of values, discourses and institutional practices in the (historical and ongoing) validation of climate change, in order to explore the ways in which science, and the epistemologies of knowledge it supports, has proven problematic for the definition and communication of climate change. The chapter firstly offers a short history of the development of climate science (see Hulme 2009 for an extended history), in order to illustrate the definitional struggles between science and politics in the identification and validation of human induced climate change. It then moves on to explore how the epistemologies of science conceptualise nature, vision and time, and the problems these conceptions have posed for understanding and addressing the complexities of contemporary environmental issues. In this context, the chapter also examines how other social and political actors, in the form of environmental NGOs and the mass media, have reproduced and reinforced these conceptualisations through their representational and communication practices. In doing so, the chapter calls for a dissolution of historically dominant conceptions of the environment as a visible nature separate from humans, to a more complex understanding of the environment in which humans, and their social, economic, cultural and political systems, are embedded (Latour 2004). As such, this first chapter provides an historical and theoretical framework which identifies the conceptual and representational practices that have impeded our understanding of climate change, and affected our ability to take action. How, and the extent to which, these conceptualisations are reproduced and/or reconfigured in contemporary climate communication and action will be the focus of the second section of the book.

The history and politics of climate science – uncertainty, prediction and consensus

As many (social) scientists have argued, climate change has been reliant upon science and technology for its detection, and for predicting the various scenarios of its future development and impacts (Weart 2003; Adam 1998a; Beck 1992). As early as 1979, the WMO organised the first World Climate Conference to address concerns about the effects of human activity upon the climate and to call upon the world's nations 'to *foresee* and to *prevent potential* man-made changes in climate that *might* be adverse to the well being of humanity' (IPCC 2004, 2, my emphasis). Spencer R. Weart traces scientific concern about human influences on the world's climate back further, to the late nineteenth century, through the work of the Swedish scientist, Svante Arrhenius, who calculated that a doubling of CO2 in the atmosphere could raise global temperatures by 5/6°C (Weart 2003). The dominant scientific view up to the first part of the twentieth century, however, was that climate was essentially stable (ibid).

The physical properties of climate are defined through meteorological statistics comprised of numerical data. In the late nineteenth century, meteorologists produced these standardised methods of observation and measurement to measure climate and weather, which involved 'quantifying it locally at individual places and, subsequently, by constructing statistically aggregated climates from geographically dispersed sites' (Hulme, et al. 2009, 197). Established in 1950, the WMO defined climate through the statistical methods established in the late nineteenth century. In order to ascertain climate, they identified (and still employ) a 30-year time period as the baseline for defining the climate of particular regions (ibid). From the basis of 30 years of accumulated meteorological data, future predictions about climate are then made. In the 1950s, however, climate was still perceived as a stable entity, and it was not until 1965, at the 'Causes of Climate Change' Conference in Colorado, USA, that scientists agreed that climate was neither simple nor stable (Weart 2003). In doing so, the scientific orthodoxy that changes in climate occurred gradually over thousands of years was being questioned, as the potential for climatic changes within a far shorter period of time were now being considered as very possible. Computers began to be employed to generate statistical climate models and to explore potential climate variations (ibid).

During the 1970s, alongside the emergence of the environmental movement, climate experts believed there would be a future cooling period rather than a warming. By the end of that decade, however, the scientific consensus was that the climate would warm as a result of increased greenhouse gas emissions (ibid). Yet, because there was evidence and experience of a cooling period in the northern hemisphere between 1940s and 1960s, it was very difficult for the public and politicians to believe that global warming was happening (ibid); a powerful and early indication that public and political acceptance of climate science is dependent upon the need for climate to be made culturally meaningful (Hulme 2008, 2009), and made evident in the present rather than the future.

According to Weart (2003), up until the beginning of the 1980s, data generated from thousands of weather stations across the globe did not appear to indicate a common standard with regards to warming or cooling trends. At the beginning of the 1980s, two scientific groups, one British, one American, undertook the task of separately examining the data in more detail to try to ascertain global temperature trends. Both groups identified a global increase in temperature, and by the mid 1980s scientists were agreed that global temperatures had increased over the last century, and that CO2 and other greenhouse gases had also risen during this period. Yet, achieving scientific consensus is never simply about producing the research data. Climate scientists required institutional support in order to gain credibility both within and beyond the scientific community and to avoid further fragmentation of their research findings. At the time, climate science was comprised mainly of the work of meteorologists and oceanographers. Although the WMO had established the first World Climate Conference in 1979, more national and international collaboration and funding was needed for the legitimacy and developments of climate science. As scientists began to collaborate across disciplines, international meetings were set up through the 1980s, and attempts were made to persuade politicians about the problem of climate change and secure further research funding.

1988 is considered a significant year for politicisation of the climate issue, and for garnering media and public attention. A series of unprecedented heat waves and droughts in the USA provided an associative link to the issue of global warming, which was covered extensively in the USA news media (Weart 2003). In the UK, the Prime Minister, Margaret Thatcher, delivered a speech to the Royal Society in which she declared global warming to be a key issue, placing this issue on the political agenda. This was accompanied by a significant increase in news media coverage (Farrow 2000; Weart 2003; Carvalho and Burgess 2005). Given the relative underfunding and low profile of climate science within the scientific community and to (inter)national governments, the sudden politicisation of climate change meant that scientific research was afforded a higher public, political and scientific profile. In reality, however, 'there were still only a few hundred people in the world who devoted themselves full-time to the study of climate change' (Weart 2003, 157).

The establishment of the IPCC in 1988 illustrates the institutional and ideological interweaving of science and politics. Conducting no new research, the role of the IPCC is to assess existing scientific reports published by the international scientific community in order to provide a comprehensive overview of the status of climate science and consensus on human induced climate change. In order to provide policy recommendations, the IPCC are also tasked with addressing the question of impacts, mitigation, adaption and the socio-economic dimensions of climate change. Comprised of scientists from a broad range of countries and a number of scientific disciplines, the IPCC has come to constitute the most authoritative scientific voice on the causes, impacts and effects of global climate change (Houghton 2004). Yet, it is not just scientists, but representatives of national governments, who constitute membership of the IPCC. Whilst this provides a crucial link between climate science and policy, it has also led to political pressure from the delegates of national governments to alter the final wording, and thus the knowledge claims, of the assessment reports. Publication of the IPCC's first assessment report in 1990 was the result therefore not only of the accumulative work of hundreds of scientists working on climate around the world, but also the political interventions of government's seeking to moderate some of the statements made. The final report was 'not mainstream science so much as conservative, lowest-common-denominator science' (Weart 2003, 162). As a result, the scientific consensus represented in the IPCC assessment reports (published in 1990, 1995, 2001 and 2007 respectively) is cautious in its judgement.

Having achieved a high degree of media coverage and public interest in 1988, climate change went off the political and media agenda for a relatively long period of time (Carvalho and Burgess 2005; Boykoff and Boykoff 2004). Yet, the work of climate scientists and the IPCC continued, and their second assessment report in 1995 was the first official international declaration that human activity was affecting global climate. By the time of the publication of the third assessment report in 2001, 'the existence of a climate response to human activities and the sign of the response' were identified as robust findings, whilst 'key uncertainties' were 'concerned with the quantification of the magnitude and/or timing of the response', as well as 'model projections' of future emissions of greenhouse gases and changes to global climate (IPCC 2001a, 30). In other words, the observable signs of the existence of climate change were undeniable, whilst the exact magnitude of change as well its effects were less quantifiable.

The credibility the IPCC has achieved as a result of the painstaking process of drafting reports and statements which have to be agreed by all scientists and government officials comes at a price. Beholden to political power, the threat of climate change is downplayed. As a consequence, uncertainties related to the nature and magnitude of climatic changes have been misinterpreted by the media and the public as evidence of the uncertainties of climate science, making it easier for powerful corporations to undermine the credibility of climate science through PR and advertising. Even the 2007 IPCC Assessment Report, which stated unequivocally that the climate system was warming as a result of greenhouse gas emissions in the atmosphere, did not escape the lobbying powers of corporations. Funded by ExxonMobil, the think-tank American Enterprise Institute (AEI), offered cash to economists and scientists to dispute the scientific claims made in the report (Sample 2007). None did.

Science and the environment

The historical developments of climate science illustrate the interweaving of science and politics in the legitimation of scientific knowledge, and the political contestations over the strength of scientific claims about the existence and magnitude of climate change. The uncertainties which have characterised the ability of scientists to make absolute claims about climate change have been used by businesses and governments to undermine scientific credibility. Part of the reason this has been successful is that science is perceived by society as a praxis based on objectivity and materiality. Yet, as an 'industrially produced phenomena', climate change is 'characterised by invisibility and periods of latency' (Adam 1998a, 10). Climate scientists have had to make future predictions about climate change based upon existing data. Yet, normative conceptions of science do not allow for uncertainty, making it difficult for climate scientists to legitimate and communicate their findings to governments, publics and the media.

The nature of climate change means that it is impossible to define exactly how rapidly, or the ways in which, climate systems will change. Climate is unstable, and does not adhere neatly to the rationalising logic of science. This instability has posed a paradox for climate scientists and for the legitimation of their research; working with uncertainties and future predictions does not fit with the empirical 'truths' demanded of classical science. Barbara Adam notes that for modern industrialising societies, science 'is the knowledge system that is called upon by the globally powerful to establish the truth, provide proof and guide public as well as political action' (Adam 1998b, 227). The IPCC fits with the model of an authoritative international scientific institution, given the task of corroborating the scientific consensus on climate change and making policy recommendations. Yet, as Newell points out in relation to the IPCC, 'By seeking to validate arguments by reference to science, policy-makers allow the scope of future options to be partly constrained by the evolution of scientific research and ensure an ongoing demand for scientific advice' (Newell 2000, 41). We have also already seen how the knowledge produced by this particular institution has been moderated and constrained by political pressures, illustrating how the production of scientific knowledge on climate change, in this instance, is circumscribed by the processes of political mediation (Shackley and Wynne 1996). Furthermore, as an international organisation which advocates international climate policy, the IPCC has faced criticism for inscribing a northern and western perspective through its focus upon policy measures that are about economic efficiency, rather than human wellbeing.

Scientists working on climate change are representative of what Val Plumwood calls the 'double face of science' (Plumwood 2002, 39). Often hailed 'as the ally and saviour of the environment' for 'exposing environmental damage and aiding opposition to it', modern techno-science has also 'contributed to producing the environmental crisis as much as to curing it' (Plumwood 2002, 39) As a product of industrialising, capitalist societies, modern science is meant to be 'value neutral', 'objective' and 'true' (Adam 1998a, 27); called upon to identify and validate environmental issues. The very notion of objectivity, however, is representative of a discourse of 'economic rationalism and productivism' (Plumwood 2002, 39). The productivist rationale of science, with its focus upon economic and technological development as progress, has contributed to the modern industrialising hazards that we are presented with today. Climate change is the result of the massive increase in the burning of fossil fuels for the development and sustenance of capitalist societies since the industrial revolution. As science has presented nature and natural resources as commodified objects to be utilised for economic production and growth (Plumwood 2002), science is thus implicated in the very development of climate change.

This duality of science, as both detector of, and contributor to, contemporary environmental risks, has been most notably examined by Ulrich Beck through his risk society thesis (Beck 1992). Beck (and many others since), has argued that contemporary environmental risks, such as climate change, radioactivity, chemical hazards, genetically modified foods, toxins and pollutants, are the products of modern industrial societies. What characterises such risks is that they are global in reach, often invisible to the senses, incalculable, and long term (Beck 1992). Involved in the very production of environmental risks, scientists, however, are still called upon to define and legitimise these risks. As Beck says, such risks 'generally remain invisible, are based on causal interpretations, and thus initially only exist in terms of the (scientific or anti-scientific) knowledge about them' (Beck 1992, 23). Because the risks (of climate change) are often 'neither visible or perceptible', they 'require the "sensory organs" of science – theories, experiments, measuring instruments – in order to become visible or interpretable as hazards at all' (Beck 1992, 27). Beck's thesis is supported by the primary role of climate scientists, climate modelling and the IPCC in the validation of climate change (Shackley and Wynne 1996; Budescu et al. 2009). However, as both producers and definers of climate change – the 'double face of science' (Plumwood 2002, 39) – the logic and legitimacy of scientific epistemologies and methods are called into question.

Although science is authorised to define the risks within a risk society, at the same time, science is itself both constrained by its own concepts and methods, and its expertise is increasingly open to social critique. Regarding environmental risks as those of nature, rather than humans, science ignores 'matters of social and cultural significance' (Beck 1992, 24). Furthermore, because risks such as climate change are not simply definable through recourse to quantifiable facts, but require both understanding and making meaningful through social, cultural and political norms and values (Hulme 2009; Hulme et al. 2009), then climate change risks become open to 'definitional struggles' by the competing interests and claims of those different groups (already) affected, as well as the 'various agents of modernity' (Beck 1992, 29), such as industries and governments. '[T]he sciences' monopoly on rationality is broken' (Beck 1992, 29) as other social and political actors, like the mass media and environmentalists, become key players in the definition and construction of climate change risk.

It is in the struggle for definition and meaning where the question of the symbolic, cultural, ethical and social dimensions of climate change becomes important. Part of how climate change has been made meaningful, and the limitations of these meanings, is related to how dominant conceptions of scientific and environmental knowledge are interrelated. I will now move on to explore some of the characteristics which underpin scientific knowledge in relation to how the environment is figured, in order to examine how these conceptions have been reproduced beyond science, both in the representational politics of the environmental movement, and in media coverage of climate change over time.

Conceptualising the environment – nature, vision and time

Nature and vision

One of the difficulties the natural sciences have faced in their historical struggles to identify climate change and legitimate climate science has been the way in which the environment has been conceptualised within the epistemologies and discursive practices of science. Coterminous with the developments of modern western science and the emergence of a capitalist society in the late eighteenth century, has been the figuration of the environment as 'nature' within scientific discourse and practice. Underpinning this figuration of the environment as nature is the separation of nature from culture, where the natural world is conceived as the non-human. Such a separation presents nature as a series of externalised objects, separate from humans and their social values/structures. As a series of objects, nature can be made knowable through the materialist praxis of science, and utilised by humans. The idea of nature and landscape as the material non-human world has been linked to the developments of science in the sixteenth and seventeenth centuries, and reinforced further through the emerging capitalisms of eighteenth and nineteenth century Europe (Wilson 1992). Capitalist societies commodify nature by presenting it as property and product, to be bought, owned and sold by humans. As nature is assigned an economic value within a capitalist society, science is authorised to provide knowledge about this nature through 'the ideology of objectivity' (Plumwood 2002, 42). The dualisms of nature/culture, subject/object, human/animal and person/property constitute the conceptual foundations of western science and capitalist societies, and thus its view of the environment.

Conceiving the environment in this way has made it difficult to identify climate change, because climate is not always visible, easily identifiable, nor an object able to be bought and sold.[1] Furthermore, the economic model of nature as object has contributed to the very production of greenhouse gas emissions and the development of climate change as a result of the extraction, use and ownership of the world's natural resources, in the form of oil, coal and gas. The scientific ideology of objectivity, which distances the self/subject from other/object, has also been criticised by feminists for reproducing gendered power relations. Plumwood observes that 'objectivity is usually seen as excluding the emotional the bodily, the particular, the personal, and of course especially the "political"' (Plumwood

1 The EU Commission's emissions trading scheme is an example of how an economic model is being used to address climate change (European Commission 2005), but one that is criticised for actually leading to an increase in greenhouse gas emissions.

2002, 42). The distancing of self/subject from other/object as a condition of the rationalising and objective logic of scientific knowledge represents nature, the body and emotion as feminised spaces, to be controlled through the 'mastery' of science and technology. The belief that climate change can be controlled by technology is an effect of both the gendering of nature and of science, but one that fails to understand how such conceptions have helped create climate change.

Central to the depiction of the environment as the natural world separate from humans and culture – a nature able to be bought, sold and controlled through scientific and economic rationality – is the role of vision, or the visual. The material praxis of science places vision and sight as the primary sense for achieving objectivity and rationality. The centrality of vision to scientific epistemology manifested in the visual taxonomies of the natural world produced by Linnaeus in the early eighteenth century (Macnaghten and Urry 1998), which also inscribed a gendered view of nature (Jordanova 1989). The imbricated developments of scientific knowledge as the objective material world that can be seen, and an understanding of nature as both visible and separate from humans, have placed sight/seeing as central to scientific and environmental knowledge. As a development of eighteenth century science, the environment understood as visible nature places knowledge in the observable structures of the world and not in its invisible relationships and functions (Foucault 1970; Macnaghten and Urry 1998). This has presented profound difficulties for identifying, validating and communicating climate change through a scientific and cultural value system that authenticates knowledge and truth through the visual.

Western science's construction of a visible nature has been supported through its visual representation within science, and in wider cultural discourses and practices. Through scientific illustrations, landscape painting, photography and, since the 1960s, satellite images of the earth from space (Ingold 1993), representations of nature have inscribed a visual epistemology through a visual aesthetics. The visible beauty of nature is captured through observation and representation, where seeing becomes metonymic for understanding and valuing this nature. NASA's online Earth Observatory (NASA 2009) – an image catalogue of satellite images of the earth – reinforces this visual spectacle of nature. Via its 'Image of the day' and its special 'NASA Global Climate Change' collection, the environment is constructed as a visible compendium of knowledge, where satellite images of a climatically changing earth work to aesthetically transfix and disengage the viewer (ibid). Aestheticising nature as a means of valuing it further externalises the environment, offering it up as a form of visual consumption, either through 'capturing' this beauty in the form of pictorial representation, or through the practices of tourism and the 'tourist gaze' (Macnaghten and Urry 1998). Visual images of nature present knowledge through the visual senses, leading to a disembodied and disconnected view of nature/environment as something 'out there'. Although tourism may involve a physical experience of being 'in' nature, such experiences have already been (mass)mediated by images of nature, whose beauty is simultaneously signified and experienced by that which can be seen. The visual disembodiment invited by tourism is manifest in its most recent developments, as westerners are encouraged by the media to go and *see* places like the Maldives and the polar ice caps before they are lost to the impacts of climate change. The media's identification and promotion of '30 places you need to visit while you still can' (The Independent on Sunday 2007) elides the contribution of travel to greenhouse gas emissions by unethically authorising the touristic practices of visual consumption.

The modern environmental movement has also contributed significantly to the visualisation of nature as indicative of the environment. The first satellite images of the earth from space have been attributed to helping the development of modern environmentalism, instigating a view of the world as a single, self-regulating organism, but simultaneously fragile and in need of protection (Cosgrove 1994; Franklin et. al 2000). The three largest international environmental campaign groups who began in the 1960s/early 1970s – WWF (1961), Greenpeace (1971) and Friends of the Earth (1971) – relied upon photographic images of a beautiful nature at risk to coralle support for their conservation and environmental efforts. Using images of nature drew upon the existing iconography of conservationism, the focus of which was upon the preservation of animals, plants and their habitats. WWF encapsulated a conservationist approach that had developed in the USA during the late nineteenth century, but brought it to an international level. Greenpeace and Friends of the Earth adopted a wider remit than simply the conservation of animal and plants species and their habitats, addressing issues of industrial pollution, ecology and human rights, as well as deploying direct action. However, they still used images of a beautiful nature at risk from humans, serving to reinforce the distinctions between nature/culture and human/animal. Greenpeace has most obviously invested in the power of the image, not only through the photographic documentation of nature and its destruction, reproduced as part of its campaign materials, but also through the visual tactics of its protests, designed to capture the attention of the news media (DeLuca 1999; Hansen 2000; Doyle 2007, 2009a; Cottle 2008). While international environmental groups have undoubtedly played a crucial role in helping to draw public, political and media attention to environmental issues, particularly climate change (Greenpeace 1990, Doyle 2007), they have also problematically subscribed to the view of environment as nature through adherence to the visual as a form of environmental documentation and persuasion.

Historically, the three most important actors involved in the communication of climate change have been scientists, environmentalists and the media (Mormont and Dasnoy 1995). For all three actors, the visual is central to the production of knowledge and the circulation of cultural meaning. Trumbo comments, 'Historically, images have always been part of science' (Trumbo, J. 2000, 380). For scientists, this may be through direct observation of empirical data, through imaging technologies such as telescopes, microscopes and x-rays, or through computerised visual simulations such as climate modelling. Indeed, climate scientists are dependent upon computer model simulations to predict unseen climatic changes in the future. The visual rhetoric employed by environmentalists (DeLuca 1999; Hansen 1993; Doyle 2007, 2009a) has been attributed to the crucial role played by the mass media in the communication of environmental issues since the early 1970s. The visual nature of mass mediated communication, particularly television (Shanahan and McComas 1999; Cottle and Lester 2009) and now increasingly newspapers and the internet, help construct and reproduce specific representational conditions, identified as dramatic, spectacular and events based news coverage that requires visual imagery. In a mediated world where 'connectivity, saturation and immediacy' are the conditions of experience, 'the establishment of liveness' has become the 'mode underlying the way in which our world is constructed as "news" by and through television' (Hoskins and O'Laughlin 2009, 3, 25). Research has shown that people's perceptions of climate change tend primarily 'to be based on photographic or video images' (Nicholson-Cole 2005, 264) drawn from television and newspapers. In other words, images which signify culturally as indexical representations of 'reality', rather than the predictive simulations of computer modelling, have dominated people's perception of climate change.

Mediated through photographic and video images, public perceptions of climate change are culturally specific and overwhelmingly negative. Anthony Leiserowitz has found that for the American public, the dominant image associated with climate change is 'melting glaciers and polar ice' (Leiserowitz 2006, 56), followed by impacts to non-human nature. In the UK, Lorenzoni et al. (2006) found that the public's most frequent association of climate change was related to the weather, and that overall, perceptions were also predominantly negative. For both the UK and USA public, climate change was not considered personally relevant, as 'the impacts of climate change, and most importantly its causes and solutions, were psychologically distant for most individuals in both nations' (Lorenzoni et. al 2006, 278). More recent work has further highlighted that the UK public perceive climate change 'as a distant issue in both time and space' (O'Neill and Nicholson-Cole 2009, 15), with public perceptions focused upon 'large-scale impacts' not considered to be personally relevant. Although participants cited 'dramatic visions', including human and animal suffering 'at both local and global scales', as making climate change 'seem most important' (ibid, 17, 18), such images concurrently worked as a form of personal distancing and disempowerment.

As photographic and video practices are dependent upon capturing visual evidence in the present to gain their representational legitimacy, this poses difficulties for rendering climate change as a present reality because of its periods of invisibility and latency. As the 2007 IPCC report stated, 'even if greenhouse gas concentrations were to be stabilised', 'anthropogenic warming and sea level rise would continue for centuries due to the time scales associated with climate processes and feedbacks' (IPCC 2007a, 16). Thus, climatic changes in the future have already been committed to in the present, but this present remains invisible, making it difficult to communicate as a reality without visual evidence. As such, climate change renders problematic not only the ways in which the environment has been constructed by science and environmentalism as a visible nature, but also the limitations of visual technologies and practices called upon to give it a sense of reality in the present and future.

Nature, vision and time

Visualising nature as representative of the environment limits the ability to identify and communicate environmental issues that are not always visible. The difficulties lie not only in the truth discourse of science, dependent upon visible materiality and reinforced through cultural investment in images of nature, but also in the temporality of environmental issues. Due to the time lags in environmental feedbacks, it can take 30-40 years for the effects of climate change to materialise (Hale 2008). Such 'periods of latency' (Adam 1998a, 10) which characterise the complex temporality of climate change, are at odds with the focus upon the visible present within scientific knowledge and in popular cultural discourses of the environment. A 'temporal disjuncture' (Adam 1998a, 34) occurs between scientific praxis which authenticates knowledge through reference to the visible present, and the effects of climate change which develop and accumulate, often invisibly and neither temporally or spatially consistent, over time.

Barbara Adam argues that organic environmental time is rhythmic, cyclical and interconnected. In contrast, clock/machine time fixes time, making it singular, finite and isolated from the processes of life (Adam 1994). Clock time, as a development of Newtonian science, seeks to make time measurable, quantifiable, predictable and commodified. It does so by measuring motion: 'when something moves its covers distance which takes time. It is time taken, the measured duration between frozen events, the mathematical statement of acceleration and slowing down, of rates of change, of the difference between before-and-after measurements of fixed states' (Adam 1998a, 40). Adam calls this atemporal time, because it 'is indifferent to spatio-temporal locations and measures time by 'extracting parts from their interactive, interconnected wholes, seeking the unit-part in its ultimate simplicity' (ibid). Because motion and time are perceived as measurable, they are also perceived as reversible. As Adam points out, however, this assumption of measurability and reversibility can only work,

> if one excludes friction, if one excludes gravity, if one excludes interaction, if one excludes life, if one excludes knowledge, if one excludes any kind of human activity, emotion, interest and frailty, if one excludes all that, then one is left with a universe of perfect symmetry, single parts in predictable, uniform motion and reversibility (ibid).

Clock time thus seeks to quantify and standardise time through decontexualisation, where an hour is the same whether it is at night or day, in the UK or Brazil, and for every human being, regardless of cultural or geographical specificity.

Technologised, standardised time measures time in such a way as to remove it from the complex rhythms and contingencies of life. While clock time has enabled the identification of changes to the climate over a period of time through the accumulation of meteorological data, it has simultaneously locked climate change into scientific definitions and predictions about the rapidity and magnitude of future climatic changes. Because clock time gives an impression of order, certainty and controllability, future climate change is assumed to be controllable, hence reversible (e.g., 'we have 5/10/15 years to tackle it'). Yet, because clock time cannot with certainty predict the exact nature or rate of future climatic changes, such uncertainty may, at the same time, be taken as the basis for inactivity in the present (e.g., 'As the future is uncertain, we won't do anything until it is certain'). This simultaneous notion of control and uncertainty, inscribed through the scientific conception of time, makes it difficult to understand the complex temporality of climate change, where changes can occur suddenly, or progressively, accelerating or decelerating at different times, with different effects, in different geographical places.

The assumption of reversibility embedded within clock time has particular resonance for climate change. Although the reports of the IPCC do not argue for the reversibility of climate change, their use of clock time to predict climate change through years, decades and centuries, unwittingly offers the spectre of reversibility through the very use of standardised, technologised time. The use of technology to quantify time, and capture numerical data, instils an investment and belief in the application of technology to 'fix' climate change. Technological fixes, from carbon capture and storage (CCS), nuclear power and 'clean' coal, to the more absurd proposals to artificially cool 'the global climate by releasing particles of sulphur in the upper atmosphere, which would reflect sunlight and heat back into space' (Connor 2006), all subscribe to this techno view, as the objective and rational approaches of techno-science are presented as the means of controlling this unruly nature. When we consider climate change from the perspectives of environmental time, however, such investment in the power of technology is revealed as a myth and a fallacy.

Clock time is also dependent upon linear perspective, the characteristics of which are 'abstraction from context, objective observation, quantification of sense data and the single fixed focus' all of which 'constitute the bedrock upon which the laws of traditional science are built' (Adam 1998a, 39). Clock time and linear perspective are the conceptual tools of a scientific view of the world that seeks to map space and time through a standardised and 'universally applicable' (ibid, 41) system. Environmental time – the multiple, complex, contextual rhythms and processes of ecosystems – is thus in conflict with the economic and scientific knowledge systems of industrial time which seek to produce objective and rational forms of knowledge. A linear notion of time and space fails to recognise the ways in which our actions do not always produce immediate or visible effects. Climate change impacts are bound in neither time nor space becoming visible only 'after they have combined to form a critical mass' (ibid, 34), rendering problematic the conceptual tools of science called upon to identify, validate and provide solutions to climate change.

Climate scientists and the IPCC have been dependent upon clock time to make predictions about the future rate and magnitude of climatic change, based upon the accumulation and examination of existing meteorological data. As Hulme et al.

(2009) have noted, the WMO (as the main scientific body for meteorology and the organisation that helped establish the IPCC) identify a baseline of 30 years from which to make 'credible' observations and predictions about climate and its variations. The 30-year time period is defined as 'normal' climate. 'Normalising' climate means 'attaching period durations to aggregated weather statistics for the description of climate' (Hulme et al. 2009, 199). Thus, '[t]hrough statistical normalisation, climate becomes enumerated and universalised' (ibid, 200). Yet, climate is not only physical it is also cultural, in the way that it is perceived and experienced by individuals and social collectives. From an understanding of climate as both physical and cultural, Hulme et al. question the appropriateness of the WMO's choice of a 30-year baseline to define 'normal' climate, as opposed to a baseline of 10 years, for example. A 30-year baseline, they argue, fails to take into account the socio-cultural dimensions of climate, in relation to individual and collective memories of weather/climate which provide a reference point to help make climate change meaningful in the present. A shorter time period of ten years may make climate change more socially relevant to individuals and a social collective because memory of climate over a 30-year span may be too long to be made meaningful to younger generations (ibid). Time, as constructed by science, and used to make observations and predictions about climate change, is universalised in a way that makes it difficult to make meaningful at a local level.

The discourse of scientific truth thus functions through an epistemological framework that structures time, space and vision as objective, mechanical and rational. The certainties of acquiring scientific truth through visible materiality are in conflict with the uncertainties that characterise climate change in terms of its specific effects in particular locations, and its rate of change. Western culture's dependency upon science and technology to validate environmental risks is endemic to the ways in which environmental issues are defined in the first place. Dependent upon science for validation, at the same time, the scientific conceptions of nature, vision and time, called upon to make climate change real and meaningful, constrain the ability to understand its complexity. Yet these conceptions are not limited to the domain of science, but get reinforced through wider cultural practices. I have already briefly shown how the environmental movement has unwittingly contributed to a conception of the environment as visible nature, and how this is also reinforced through mass mediated communication. I want to finish this chapter by very briefly examining the role of the media more specifically within climate change communication (explored in more detail in Chapters 2 and 5) in order to illustrate how scientific conceptions of nature, vision and time get reinforced through mass mediation, but also how the notion of scientific uncertainty, discussed at the beginning of the chapter, is reinforced through the conventions of news coverage.

Media coverage of climate change

The media's engagement with climate change has been characterised by 'a long period of passivity punctured by a more recent burst of alarmism' (Andreadis and Smith 2007). Although media coverage of climate change across the globe has increased since the late 1980s, with a more substantial increase from the early 2000s as impacts of climate became more visible (Boykoff and Roberts 2007), there has been a noticeable time gap between the establishment of scientific consensus on human induced climate change in the late 1980s/early 1990s and a sustained engagement with this issue by the world's media. The media's lack of interest in climate change has been shaped, in part, by the limitations which underpin science itself, circumscribed by the specific context of news conventions. Journalistic norms of scepticism have undermined the consensus of climate change scientists (Andreadis and Smith 2007). This has been further affected by the notion of a balanced argument, leading to what Boykoff and Boykoff (2004) have called the 'balance of bias' in media coverage of climate change, as the small minority of climate sceptics (often funded by powerful industry lobbies) have historically been given equal coverage as climate scientists. Added to this has been the complexity of the issue, which demands certain expertise from journalists, often lacking in media organisations (Andreadis and Smith 2007).

Concepts of nature, vision and time structure the production of knowledge about the environment and the ways in which knowledge is given meaning. Nature/culture distinctions which have dominated western conceptions of the environment, and which privilege visible materiality as a condition of scientific knowledge, have also impacted negatively upon media coverage of climate change. Until fairly recently, there has been a relative lack of images (both still and moving) of climate change. This lack of a visual presence has not only affected public belief in climate change, but also contributed to a lack of media coverage. In 1990, a former BBC news correspondent commented: 'Above all, environmental stories need good pictures…global warming is actually very difficult because you can't actually see global warming' (cited in Anderson 1997, 121). As a consequence of the categorisation of the environment as a visible nature, supported by a discourse of visual truth, the lack of a visual dimension to climate change has affected news coverage.

As images are used to authenticate a visible present, then the relationship between vision and industrial clock time has also shaped media communication on climate change. How time and vision are imbricated in the production of environmental knowledge is revealed by a former ITN news environment correspondent in 1993, who commented, 'Pictorially it's a tricky one to show global warming because they're showing something of the future' (cited in Anderson 1997, 122). Scientific conceptualisations influence media discourses and practices, making it difficult to make climate change credible and meaningful. While the visual spectacle of climate that currently characterises news discourse (Cottle 2009) certainly functions to make climate more 'real', this focus upon the spectacle is still constrained by the same representational limitations. The focus upon dramatic climate spectacle reproduces and reinforces the concept of a visible nature at risk, leading to a sense of fear, disconnection and disempowerment on the part of the viewer (O'Neill and Nicholson-Cole 2009), whilst subscribing to an epistemology of visual 'truth'. The long term nature of climate change as an ongoing and developmental issue has been further constrained by the shorter time scales and limitations of daily newscycles.

Media communication on climate change also depends upon the use of symbolic or 'cultural resonances' in order to make the issue meaningful. More so than other environmental issues, the complex nature of climate change has made it difficult to 'link into powerful, historically established, symbolic imagery' (Hansen 1991, 451). Anderson similarly points out that certain issues gain public legitimacy through their capacity to signify symbolically; '(t)hese issues tend to have particular "carrying capacities", they become icons or symbols for a wide range of concerns that people can easily identify with' (Anderson 1997, 5-6). Shelly Ungar's (2000) analysis of the very different mobilisations of media and public attention regarding the issues of climate change and the ozone hole are a pertinent example of how symbolic and cultural resonances impact upon media and public engagement with environmental issues. Ungar argues that the ozone hole was 'allied and resonated with easy-to-understand bridging metaphors derived from popular culture', such as the 'penetration metaphor' (Ungar 2000, 298, 305). Here, the notion of rays bombarding the earth and penetrating a shield resonates with images and narratives from popular science fiction, Hollywood films and video games. In contrast, climate change cannot invoke such a simple and direct metaphor because it is a more complex issue, thus 'creating the problem of finding the human "fingerprint" amidst highly variable and complex natural processes' (Ungar 2000, 305). The range of climate change images we now see in the media are in some sense now able to also draw upon existing symbolic resonances regarding the landscape, nature and animals; or nature/culture distinctions. Unlike the ozone hole, however, climate change mitigation has not achieved the same level of public, political or media engagement, and cannot be so simply addressed.

Conclusion – moving beyond nature

This chapter has examined the historical developments of climate science and the conceptualisations which underpin scientific knowledge of the environment. In doing so, it has called attention to the definitional struggles between science and politics in the construction of climate change knowledge, identifying how notions of scientific certainty and uncertainty have been used politically to undermine climate science. At the same time, however, the chapter has argued that the concept of certainty within classical notions of science has been called into question by the very nature of climate change and the practices of prediction, forecasting and modelling which underpin climate science. Crucially, how modern science conceives of the environment as a feminised, commodified, disembodied and visible nature, quantifiable through numerical measurements based upon a universalised system of time, has proven problematic not only for the identification of climate change, but also the ways in which it is given meaning, and thus the forms of technological action advocated. Latour's observation that 'the end of nature is also the end of a certain type of scientific certainty about nature' (Latour 2004, 63) is particularly pertinent here.

Scientific conceptions of (environmental) knowledge are inscribed and reproduced through wider cultural discourses and practices. The mediating practices of both the environmental movement and the mass media have contributed to the representation of climate change as a nature 'out there', rather than an integral, embodied part of human culture, inscribed by social norms and values. As climate change fundamentally questions scientific values and current ways of understanding the environment, then a new way of thinking about the environment and human relationship is necessary; one which situates people's everyday interactions as part of environmental processes and ecosystems. Rather than waiting for scientific certainties about the exact nature of the future impacts of climate change, which can never be certain, the feminist call for 'socially responsible and reflexive forms of knowledge...essential for democratic and ethical outcomes' (Plumwood 2000, 51) must be taken up. Although the controversy over the UEA leaked emails appears to have contributed to a decrease in UK 'public confidence that climate change is happening' (Spence et al. 2010, 17), as well as in the USA (Leiserowitz et al. 2010), the event may serve to encourage climate scientists and researchers to publically acknowledge the uncertainties which underpin climate science, and in doing so create more reflexive forms of knowledge and communication. Furthermore, understanding the power relations which underpin the conceptual distinctions between nature/culture, subject/object and self/other – distinctions which climate change breaks down – will enable a more ethical and productive engagement with the issue of climate change that is attentive to a range of different perspectives, but one that ultimately acknowledges the human costs and global inequities of climatic change, and uses these as the basis for ethical action.

Chapter 2
Visualising Climate Change: Negotiating the Temporalities of Climate Through Imagery

> [W]hatever you can *see* is not actually the global warming of the present…the warming that we see today…is a function of greenhouse gases emitted decades ago, and the full effects of the carbon dioxide we are emitting today will not be felt until decades from now…Another way to put this is that whatever you can see is the past of climate change; the *present is still invisible* (Kolbert 2009, 70, my emphasis).

As Kolbert reminds us, climate change is temporally complex and not always visible. The time lags between cause and effect pose problems for its visual representation. While historically a lack of visual 'evidence' made it difficult for scientists and environmentalists to communicate the reality of climate change to sceptical governments, publics and the world's media, the more recent proliferation of photographic and video images documenting climate change impacts (Braasch 2005; Pearce 2007; Schmidt and Wolfe 2009; NASA 2009; Cottle 2009) should not leave these difficulties unexplored. Indeed, the very concept of visual 'evidence' supports scientific knowledge systems which have figured the environment as a visible nature, making it difficult to communicate the inherent invisibility of climate change, or the temporal disjuncture between its causes and effects (Beck 1992; Adam 1998a, 1988b). Given the centrality of images to the communicative practices of scientists, environmentalists and the media, surprisingly little research has been conducted on the visual communication of climate change in comparison to textual analyses (exceptions include Nicholson-Cole 2005; Linder 2006; Doyle 2007, 2009a; Leiserowitz 2006; Lesierowitz et al. 2008; Hansen and Machin 2009; Manzo 2010; Cottle and Lester 2009; Smith and Joffe 2009). Yet, climate change provides a unique opportunity to explore both the possibilities and limitations the visual poses for the communication of temporally complex phenomena like climate change, within the context of an image-centric, western media culture.

Current analyses of climate change images have called attention to the distancing and disempowering effects of photographic and video imagery (Leiserowitz 2006; Lorenzoni et. al 2006; O'Neill and Nicholson-Cole 2009), and the reappropriation of such images, along with those of climate science, by commercial markets for financial gain (Linder 2006; Hansen and Machin 2008). If climate change imagery is increasingly distancing and decontextualised, what role have images played historically in the mediation of climate change? Given the centrality of the visual to western science and culture, how have the complex temporalities and changing visibilities of climate change been negotiated and represented through images?

This chapter explores these questions through a critical analysis of a selection of imagery produced by scientists, environmentalists and the media, from 1990 to 2007, to examine how the temporal and visual limitations of climate change have been negotiated *visually* by these social actors over time.

The focus upon scientists, environmentalists and the media is due to their historical significance in the identification of climate change and its communication to international governments and publics (Mormont and Dasnoy 1995). Comprising both a national (UK) and global context, this chapter examines IPCC assessment reports, Greenpeace (International and UK) campaign materials and BBC (British Broadcasting Corporation) television documentaries. The four IPCC assessment reports (1990, 1995, 2001 and 2007) are analysed because of their privileged status within the international politics of climate science, and their policy influence. Representing environmental communication strategies, Greenpeace has been chosen because out of all the international environmental NGOs it has been involved the longest in the communication of climate change (Leggett 1990; Greenpeace UK 1990; Mormont and Dasnoy 1995), and is specifically (but problematically) invested in the visual communication of environmental issues.[1] In the case of media representation, the visual format of television (rather than newspapers, for example) makes it ideal for the analysis of imagery; while the focus upon documentaries, over news media, addresses the relative lack of analysis of non-news factual genres within media and communications research (Mellor 2009). As a public service broadcaster, the BBC has been chosen because of its remit to inform, educate and entertain the (British) public, and its claims to impartiality and honesty (BBC 2009a; BBC 2009b). Its concern to bring 'the UK to the world and the world to the UK' (BBC 2009c) situates its communication of climate change within the context of both a national and global audience.

This chapter discursively analyses the range of images produced by these actors, at particular moments in time (synchronic), and across time (diachronic). Attentive to the context of their production and intended reception (Rose 2007), the images are also examined in relation to wider socio-cultural discourses and practices, including the status of climate science, climate policy and news media coverage at particular historical moments during the period 1990-2007. Kate Manzo's (2010) typology of climate change images provides a useful way of identifying the range of images deployed across this time period. Manzo distinguishes the categories of 'Planet', 'Environment' and 'Living Beings', and their various types: the 'Planet' is represented by globes and maps; the 'Environment' by 'flooded areas' and 'melting glaciers/polar regions'; and 'Living Beings' by 'Animals – Polar, bears, penguins' and 'People – in floodwater, on parched land, on glaciers' (Manzo 2010, 98). Significantly, people have constituted the least represented type. Alongside these categories/types of climate imagery, the analysis also considers the representational practices/technologies used, including photography,

1 My analysis of Greenpeace here draws upon, but also extends, research I have already undertaken on Greenpeace climate change communication (see Doyle 2007).

video, satellite images, scientific drawings and computer simulations. While all images are representations, signifying on a connotative or symbolic level (Barthes 1997), photography, video and satellite images also gain authority through their indexicality, or denotive function (Barthes 1977, 2000; Sontag 1977; Nichols 2001; Mellor 2009). In contrast, computer simulations are created from mathematical data, translating 'invisible entities (beyond the range of human vision)...into visible analogues' (Wolf 1999, 274). Wolf contends, however, that 'in science and in many public sectors, as well as in the public's imagination, the mathematical basis behind computer simulation has given it a status similar to (or greater than) photography, despite the often much more tenuous indexical linkage' (ibid, 286). In examining the range of imaging technologies and representational practices used in the visual mediation of climate change, the analysis pays attention to the scope of meanings generated through varying contexts of production and reception.

Drawing upon the theoretical framework presented in Chapter 1, this chapter examines how concepts of nature, vision and time have been negotiated by the three actors through visual representation. In doing so, it charts the recent historical developments of a visual lexicon of climate change within and across science, environmentalism and the media, to consider how these may have shaped our current perceptions of climate change as a distant future, rather than locally and personally relevant (O'Neill and Nicholson-Cole 2009). The analysis identifies four historical phases (presented here in four parts) in the visual communication of climate change, to explore how imagery is variously used to make climate change meaningful as a present or future issue, and to help create the discursive conditions for promoting climate (in)action.

Part 1: Predicting the unseen future through the (in)visible present (1990-1995)

Scientific visualisations – rational visions of present/future worlds

> ...we might conclude that detection [of human induced climate change] with high confidence is unlikely to occur before the year 2000 (IPCC 1990, 253).

Reliant upon science for its identification, significant international recognition of climate change came through the publication of the IPCC's first assessment report in 1990. Both the IPCC and climate science more generally have depended upon climate modelling to make future projections of climatic change. Taking the form of graphs, maps and globes, these scientific visualisations are 'speculative images' whose 'representation is based partly on theory and partly on fact', and, as such, are 'simulations based on structural prediction of processes that cannot be seen' (Trumbo, J. 2000, 382). They demonstrate the temporal complexity of climate change, where data in the present is used to make predictions about the unseen future. From a representational perspective they constitute one of the first significant sets of images of climate change to be produced, playing a crucial role in the development of climate science, and in the visual lexicon of climate change.

While computer simulations are central to climate science, Shackley and Wynne (1996) argue that the IPCC reports have failed to acknowledge the uncertainties which form the basis of climate modelling, in their efforts to influence policy decisions through the promulgation of (future) certainty. This is apparent in the 1990 IPCC report, which argues for significant investment in the production of general circulation models (GCMs) through computer technology. In 1990, three dimensional models of the atmosphere and of the ocean were produced separately, and coupled occasionally: 'even though this [coupled models] is crucial for climate change prediction, only a few models linking all the main components of the climate system in a comprehensive way have been developed' (IPCC 1990, 73). Taking into account the 'different timescales' of atmosphere and ocean requires 'computer speeds several orders of magnitude faster than is available on current machines' (ibid). Reliance upon limited computer technology is exacerbated by the different temporal dimensions of atmosphere and ocean: 'while the atmosphere reacts very rapidly to changes in its forcing (on a timescale of hours or days), the ocean reacts more slowly on timescales ranging from days (at the surface layer) to millennia in the greatest depths' (ibid, 77). The complexity and diversity of ecological time, in this instance, poses problems for the memory capacity and representational capabilities of computer technology.

Climate science, as presented by the IPCC, reinforces institutionalised discourses of techno-science through investment in the (future) capabilities of computer imaging technologies. Yet, climate modelling is dependent, not only upon the technical capacity of computers, but also 'the availability of appropriate observed data' (ibid, 100). This is a central paradox of climate modelling: improvements in computing technology and observed climate data aid the accuracy of the modelling process, yet the passage of time advances climatic change. The future capabilities of computers and 'satellite observing systems' which are 'expected to become an important new source of global data for model validation' (ibid) rely upon (imaging) technologies to generate climate data. Indeed, in 1990, the reason that increases in global mean temperature could not yet be wholly attributed to the increase in greenhouse gas emissions was because climate models were 'still in an early stage of development' (ibid, 243). Such investment in technologies is characteristic of the scientific project of global geo-research and the development of earth observational systems, instigated in the late 1960s following the production, by NASA astronauts, of some of the first photographic images of Earth from space. Wolfgang Sachs notes that 'hope of future expansion [in this case, in the production of knowledge of climate change] is directed towards observation of the earth alongside communications technology' (Sachs 1999, 125). Climate science similarly privileges the economic and power relations inscribed in scientific and governmental institutions through its commitment to the expansion of global observational projects for climate modelling.

The images produced in the first assessment report are representational; standing in for ecological processes, and utilising cultural conventions of representation in order to provide meaning. Comprising of tables, graphs and maps, the images attempt to give visual presence to invisible processes. Graphs which show ascending lines endeavour to make visible, through conventions of scientific illustration, the increase in global temperatures since 1750 and the corresponding increase in greenhouse gas emissions over the same time (produced as separate graphs in 1990). Communicated numerically by decades and centuries, the passage of time is presented as a steady horizontal line, against which the increase in global temperatures and greenhouse gases is to be measured and predicted. Time is presented as steady and ordered, and ecological processes are presented as fluctuating (see Figure 2.1), but showing a steady increase in temperature (and atmospheric concentrations of CO2/greenhouse gases) since 1750. The multiplicity of rhythms and temporalities that constitute ecological processes are thus rendered pictorially as ascending black lines. While this simplifies ecological processes in order to show climatic change and make it 'real', at the same time, the presentation of industrial time as a steady line, against which climatic changes are measured, renders ecosystems as knowable through scientific practices premised upon industrial time. The visual order imposed upon ecological processes by the graphs, which show past, present and future through ascending lines, are counter to the temporal and spatial complexity of climatic change.

Accompanying these graphs are pictorial representations of the modelled climate system attempting to link ecological processes, such as surface air temperatures, soil moisture and CO2 precipitation, to specific geographical places through the form of maps. These maps are focused predominantly on the global scale, flattening the contours of the earth into two-dimensional illustrations of the major continents

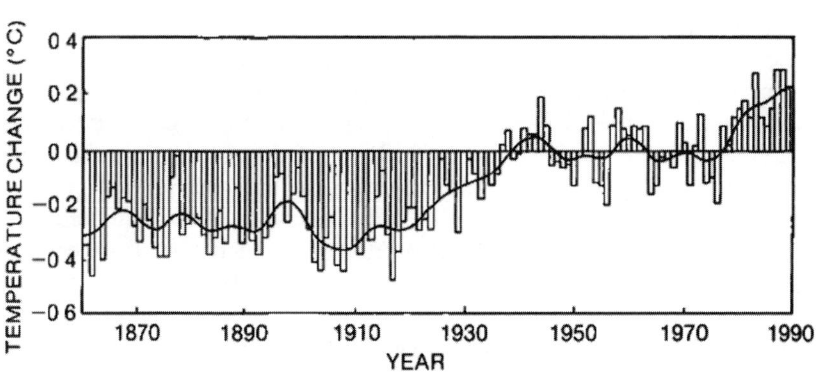

Figure 2.1 IPCC graph showing global mean combined land-air and sea-surface temperatures from 1860-1990

Source: Climate Change 1990: The IPCC Scientific Assessment Report Prepared by Working Group I, Figure 11. Cambridge University Press. Reproduced with permission of the IPCC.

Figure 2.2 Computer modelling showing effects of doubling CO2 for July-August-September on surface air temperatures (10 year means) from the IPCC

Source: Climate Change 1990: The IPCC Scientific Assessment Report Prepared by Working Group I, Figure 5.4. Cambridge University Press. Reproduced with permission of the IPCC.

of the world (see Figure 2.2). As such, they represent the world according to the cartographic and geometric rules of longitude and latitude, first used by Mercator in the sixteenth century (Whitfield 1997; Sachs 1999). Derived from computer models, the coloured maps attempt to illustrate the predicted effects in 2030 of doubling CO2 on 1990 rates. Colours associated with warmth and heat – yellow, orange and red – are used to demonstrate warming; the darker the colour, the higher the temperature. Such images are early representations of the potential global distribution of increased CO2 emissions, where the different colourings illustrate the unevenness of temperature changes. The use of the world map locates specific temperature increases to broad geographical areas, adding spatial nuances to the information presented in the accompanying graphs. Yet, the limitations of current models are acknowledged by the IPCC for failing to illustrate regional specificities.

Like graphs, the simulated maps visually construct and represent the climate system through the use of existing cultural metaphors and knowledge–power relations. As Whitfield reminds us, 'Maps are cultural artefacts' (Whitfield 1997, viii), representations of the world that construct and reflect historical, social and cultural values. Linked historically to the western colonial project, the cartography of maps invokes 'the desire to see an image of the entire world focused before us; clear, self-contained, comprehensible and masterable' (Whitfield 1997, 2). The perception of 'mastery' provided by maps is derived in part from the viewpoint they offer; representing the physical contours of the world from above. This 'space-centred perception' (Sachs 1999, 103) of the world erases a sense of place. While this representational viewpoint occurred centuries before the technical means of flight enabled it to become a 'reality', Sheila Jasanoff argues that the military and scientific endeavours of the mid-twentieth century provided the technological and ideological means to 'collect and record data on a global scale' (Jasanoff 2004, 38). The space race of the late 1950s and early 1960s initiated a new age in earth viewing, commencing with the Soviet Union's launch in 1957 of the first scientific satellite to orbit the earth (followed by the USA). This was reinforced by the International Geophysical Year (1957-8), which sought to co-ordinate the efforts of scientists from around the world in identifying global geophysical data. As such the image of the world from space was being chartered, recorded and mapped by (inter)national governmental programmes premised upon the acquisition of scientific data, and the perpetuation of knowledge–power relations within scientific and governmental institutions.

The global maps used in the IPCC report acquire some of their meaning through these historical developments. As Jasanoff observes, 'Providing facts on a planetary scale requires not only scientific instruments and work, but also institutions capable of interpreting and disseminating knowledge' (Jasanoff 2004, 45). Institutions thus 'arose to produce simulations and statistics about the Earth that could more readily serve as a basis for global policy. These bodies straddle the world of science and governance by hybridizing knowledge and power' (Jasanoff 2004, 44). As an institution, the IPCC embodies this position. The simulated world maps produced in their 1990 assessment report (and in subsequent reports) are mediations between science and politics. These maps present a dialectical vision of a climatically changing Earth: while illustrating the potential increases in global temperatures and their differential spread across the world's continents, at the same time, the use of a world map invokes a sense of control and order on climate change that is at odds with the temporal and spatial complexity of climatic changes, and the unpredictability of feedbacks.

Although the three-dimensional image of the globe is rarely used in the IPCC's first assessment report, the map of the world laid flat draws meaning from this image. Much has been written on the role of the image of the earth in the development of modern environmentalism. Produced in 1968, NASA's photographic images of Earth from space are attributed to the initiation of a global consciousness of shared humanity, representing the earth as a fragile entity under threat. The image of the earth, however, has also been critiqued precisely for instigating a global world view which erases national differences and eliminates 'people, places and political boundaries' (Jasanoff 2004, 46). Or, as Wolfgang Sachs argues, '…it stands for the unification of the world in the image of the West' (Sachs 1999, ix). As products of western science, the simulated world maps used by the IPCC draw their authority as images invested with the power of science and technology, erasing political and cultural differences. Although the simulated maps attempt to show the differential increases in temperature around broad regions of the world, they cannot show the specificities of climate change on a local level, nor do they attribute the quantity of greenhouse gas emissions produced by different countries/nations.

As one of the first sets of images of climate change, the computer modelled maps utilise existing cultural symbols and discourses to help make climate change meaningful. They attempt to show visually the ways in which increased CO2 emissions will lead to future increases in temperatures across the globe. In doing so, they are dependent upon both the capabilities and limitations of computer technology, investing in a future when computer and satellite technologies will enable better data and better simulations, simultaneously denying the uncertainties which form the basis of this science (Shackley and Wynne 1996). These simulations are constrained by the fact that they are simulations; projections of future events that rely upon observations in the present to gain their validity. Both the maps and the graphs invest seeing/sight as the basis upon which knowledge of climate change is produced, where temperature increases and changes in soil moisture are designated through broad colour coded regions on the world map. Dependent upon science for their power, at the same time, these representations decontextualise the processes of climate change from a lived and sensed experience. We look down on the earth, and observe the graphs, from a disconnected position, where the specificities of place and space are removed. Furthermore, the institutionalisation of knowledge promotes the notion of techno-fixes, deflecting attention away from moral or political obligations to address the global inequalities of climate change.

The 1990 IPCC report had been preceded by a couple of years of high media coverage of global warming/climate change, particularly in the USA, UK and New Zealand (Ungar 1992; Bell 1994; Carvalho and Burgess 2005) The issue was linked to weather related events such as drought in the USA (Ungar 1992; Krosnick et al 2000), and, incorrectly, with the ozone layer in New Zealand (Bell 1994). In 1988, media coverage was supported through political announcements on climate change by the UK Prime Minister, Margaret Thatcher (Carvalho and Burgess 2005) and New Zealand's deputy Prime Minister, Geoffrey Palmer (Bell 1994). In Western Europe and North America, however, media coverage peaked in 1990 and did not increase again until 1997 following the adoption of the Kyoto Protocol (Boykoff and Roberts 2007). In Australia, New Zealand, the Middle East, Asia, Eastern Europe and South Africa, media coverage increased slightly in 1990, remaining relatively even, but low, until 1997 (Boykoff and Roberts 2007). Media coverage of climate change in the late 1980s focused upon the occurrence of extreme weather events, such as droughts and cyclones, making the simulated futures produced by climate scientists difficult to translate into credible realities. In the context of UK media coverage, Carvalho and Burgess (2005) note that while scientists first identified the issue, they lost definitional control after 1991, as climate sceptics received more media coverage than climate scientists; a shift also reflected in the USA media (McCright and Dunlap 2003; Boykoff and Boykoff 2004).

Reinforcing climate science – televisual discourses of (ir)rationality

The visualisations produced by the IPCC have been dependent upon the media and environmental NGOs for their wider circulation. NGOs, in particular, provide 'substance to scientific diagnoses which are most often expressed in the arid forms of figures, curves and diagrams', seeking 'to define the largest number of possible consequences that may have some significance for the different publics which it addresses' (Mormont and Dasnoy 1995, 57). The BBC documentary, *The Heat is On* (BBC/OU 1991), and a Greenpeace International climate campaign report, *Emerging Impacts of Climate Change?* (Greenpeace International 1993), illustrate the different approaches taken by these actors in early attempts to the make climate change meaningful to a UK and international audience. While both draw upon scientific knowledge to explain climate change, this is framed very differently, deploying divergent sets of images to communicate the issue.

Through interviews with leading climate scientists and the prolific use of scientific visualisations in the form of graphs and climate model simulations, *The Heat is On* (BBC/OU 1991) assigns definitional control to scientists. It is a 'subjunctive' documentary because it relies upon computer imaging simulations to show 'what *could be*' or what '*would be*' (Wolf 1999, 274), but is also a science documentary, where 'the representational methods of scientific discourse invite similar expectations to those of documentary film', that is, indexicality (Mellor 2009, 143). Simulated images gain their indexical status through the 'truths' of scientific discourse, as conversely, the indexical status of video images favoured by the news media are disavowed in the opening sequence. Here, audio excerpts from TV news stories of rainstorms and floods in Britain and France are supported visually by video footage of sea storms, hurricanes and flooding. Acting as the 'voice-of-God' (Nichols 2001), the unseen narrator informs us that throughout the 1980s and into the 1990s, weather has been getting more extreme, with floods and droughts setting new records. Yet, the credibility of these visual signs is undermined by the narrator who tells us that whilst some (i.e. the news media) have claimed that these are signs of climate change, others have not. The 'others' are climate scientists, represented here by Stephen Schneider, from the National Center for Atmospheric Research (NCAR), USA. Schneider notes that in the last year, the media have picked up on global weather events, presenting these as signs of climate change. However, 'It's going to make many years to prove it' (BBC/OU 1991). Appearing to undermine the credibility of climate science, what Schneider and the documentary are doing is establishing climate science as a rational, cautious, and long term project, whose knowledge is accumulated over time. The rationality of science and the irrationality of media representations are thus presented through images, where indexicality is assigned to the future predictions of climate modelling and graphs, rather than video evidence of weather events in the present.

The contested 'truth' status of climate imagery is at the centre of the programme's exposition, as science is both textually and visually established as more credible than the indexical force of news media images. A scientific graph depicting an increase in global temperatures is directly contrasted with the visual spectacle of media coverage, as the narrator says, 'It was the slow rise in temperatures that had climatologists worried rather than extreme weather events the media were excited about' (ibid). Linguistically linking worry to climatologists and excitement to the media conveys a sense of scientific responsibility as opposed to inappropriate media hype. In this instance, scientific images in the form of climate model simulations of the future are presented as more credible than photographic/video images of the present/past (Wolf 1999).

Credibility is established further through large scale imaging and computer technologies. Accompanied by an image of a computer model of the globe, Schneider states,

> In order to make predictions you need satellites of the earth looking down to see the amount of energy that's coming in and out to see what's happening on the globe, then large computers to store, process and validate the only tools we have for forecasting the future climate models (BBC/OU 1991).

Schneider comments that computer models are 'very computer intensive', that 'we need lots of instruments', and that 'hopefully by about 2000 we'll begin to build credible forecasting skills' (ibid). In framing the issue as a concern of science and technology, the role of the (news) media in the communication and making meaningful of climate change is undermined. While the documentary's critique of the indexical status of video images helps to foreground the unseen nature of climate change, the framing of this critique constructs an inverted hierarchy of image production and type. Trust is placed in scientific and governmental institutions for the development of future technologies in order to make better climate predictions. The temporal future is thus more reliable than the present, effectively placing knowledge and power within an institutional framework.

Like the work of the IPCC, the institutionalisation of knowledge production is reinforced through the linkage of science to policy (Shackley and Wynne 1996). Alongside scientists, politicians are given precedence as the main protagonists in the co-ordination of international efforts to address climate change, effectively rendering citizen/NGO action as secondary. Video footage of politicians debating visualises the link between science and policy. In this instance, the indexical status of photographic/video evidence is not questioned, but affirmed through discourses of governmentality. Both politicians and the media, however, are criticised for demanding visible proofs. Condemning the media for distorting the debate on climate change, Schneider says, 'What gets lost in this confusion is the basic consensus', whilst what remains uncertain are 'details such as timescale' (BBC/OU 1991). Schneider's criticisms of media representations and common (mis)perceptions of scientific knowledge are supported by the narrator; 'The global warming debate has thrown into sharp focus the very nature of scientific evidence. Politicians are being asked like never before to take action based on estimates rather than established facts…science can rarely provide the 100% proof that politicians are demanding' (ibid). By acknowledging the uncertainties associated with predicting the exact magnitude and effects of global climate change within particular countries, the documentary highlights the temporal and spatial complexities of climate change. As an early example of the televisual communication of climate science to a mainstream audience, this questioning of the notion of scientific (un)certainty in the context of climate change is a timely reminder of the misrepresentation of scientific uncertainty in relation to the consensus of opinion on climate change during the 1990s (Boykoff and Boykoff 2004), but also the loss of a more open discussion of scientific uncertainty as revealed by 'Climategate' (see Chapter 5).

In foregrounding the science of climate as reliant upon computer model simulations, the documentary argues that climate change needs to be understood as long term and future based, rather than short term and visible in the present. This could be viewed as an important early critique of the spectacular nature of media coverage of climate, also highlighting the contingent nature of scientific knowledge. However, the promotion of scientific imaging processes, dependent upon substantial economic and institutional support, reinforces existing power relations, effectively disempowering the mass media, the public and NGOs. Indeed, within the institutional framework of the BBC, the documentary reinforces the discursive authority of (techno)science and government, as the viewer is called upon to place trust in these institutions. Yet, given that the programme is itself a mass media product, the tensions and contradictions inherent to its critique of media discourses are also played out. The need to make the issue culturally meaningful is revealed by the title of the programme – *The Heat is On* – which metaphorically invokes a warming planet and helps to simplify the effects of climate change for the public, reinforced by the narrator's use of the term 'global warming'. 'Climate change' is the term used by the scientific community and the IPCC, whilst 'global warming' has been the term most used by the media (Whitmarsh 2009). Interchanging between 'climate change' and 'global warming', the programme's use of both phrases illustrates the media's negotiation of scientific terminology and language that may be more culturally meaningful to a wider audience, even as it criticises the sensationalism of media representations.

Climate change in the 'here and now' – NGO communication on present/future effects

Greenpeace's climate change communication performs a more explicit motivational function. As one of the first international environmental NGOs to communicate the findings of the IPCC's first assessment report to a wider audience, Greenpeace has been at the forefront of climate change communication since the early 1990s (Leggett 1990; Greenpeace UK 1990). While its early climate publications are predominantly text based, subsequent campaign materials have integrated the visual with the textual. Unlike the cautious findings of the IPCC, and the BBC's critique of news media representations of climate change, Greenpeace's campaign report, *Emerging Impacts of Climate Change?* (Greenpeace International 1993), asserts the presence of climate impacts in the here and now. Through photographs of weather related events, such as droughts, cyclones, floods and fires, the question posed is answered in the affirmative (See Figure 2.4).

Opening with a photograph of scorched earth, accompanied by the text 'the first signs of the big heat', the report places climate change in the present through photographic 'evidence' that associates it with warming and heat. The temporality of rising heat is given immediacy through criticism of the term 'global warming': '"Global warming", the "greenhouse effect" – these are terms that do not in any way reflect the danger of what we humans are doing to the atmosphere. "Climate chaos", "climate apocalypse", the "hothouse effect" – these terms would be more appropriate' (ibid). The urgency of climate change is (re)defined through an appeal to make it linguistically more immediate, dangerous and catastrophic. Tensions emerge between the discourses of science and those of environmentalism, as the need for the climate change to be made immediately relevant to the public is in conflict with the cautious observations offered by science.

Greenpeace reappropriates the findings of the IPCC, through the inclusion of a temperature graph, citing research from other scientific bodies such as NASA and the British Scott Polar Institute. Yet, it is critical of scientists for being constrained by institutional protocols and for waiting for proof of climatic changes: 'there are *already* many worrying indicators that the climate is *already* responding to the effects of our uncontrolled emissions of greenhouse gases and that global warming is *already* being felt' (ibid, 5, my emphasis). The future predictions of science, based upon current observations, are altered to emphasise the existence

Figure 2.3 Front cover of Greenpeace's *Emerging Impacts of Climate Change?* report

Source: Greenpeace International 1993. Reproduced with permission.

and urgency of climate change in the present, reinforced by the use of active verbs to depict these ongoing changes: rising; melting; warming; dying; foundering; spreading; intensifying; shifting; shrivelling; imperilling.

As a basis for mitigatory action, this urgency is supported by visual 'evidence'. With the exception of one graph, all the images are photographs. Some depict weather 'events' that have already occurred, or show impacts to biology such as bleached corals. Working as both denotations and connotations of climate change (Barthes 1977), they show the present impacts of climate change, and make it culturally meaningful through weather. Other images in the report signify more explicitly through connotation. For example, images of animals imply the future impacts of climate change through association with the photographic 'evidence' of impacts. Importantly, future impacts are linked to both animals and humans: animals are represented by the polar bear and the Coroboree frog, whose habitats are at risk; humans are visually represented in relation to the loss of home/habitat and the return of diseases. In one photograph, a white man with a facial sore signifies the visible effects of the potential return of old diseases such as Leishmaniasis (oriental sore), malaria and dengue fever. Overall, discourses of danger and catastrophe for humans and animals prevail, illustrating how environmental NGOs 'operate a type of futurology, occasionally a little catastrophic because the message presupposes an appeal to the emotions and preoccupations and daily experiences of ordinary people' (Mormont and Dasnoy 1995, 57)

As an early example of NGO climate change communication, what is striking is the range of images used to convey the present and future impacts of climate change. The inclusion of humans is particularly noteworthy, helping to shift the environmental focus away from non-human nature, lacking in more recent visualisations (Manzo 2010). These images, however, are not exempt from critique; while they figure climate change as a global issue affecting all humans and animals, their positioning of people in developing countries reinforces a discourse of vulnerability and danger that has colonial roots (ibid). While most of the photographs in the report include an explanatory title, two images do not. One shows three black people digging in between the cracks of a drought stricken land (see Figure 1.4). The other shows two black people, laid on dusty earth, whose faces and upper bodies are covered over with sheets, signifying death. In the background are groups of black people waiting outside tents. Without any textual reference, apart from the titles of the section of the report they appear in – 'Spreading drought' and 'Shrivelling food supplies' – these images signify on a connotative level, assuming the viewers a priori interpretation of photographic images of suffering related to African famine and western aid to developing countries (Chouliaraki 2008). Intended to convey the human impacts of climate change, they are generic images that utilise an existing discourse of vulnerability that denies the western colonial roots of poverty in developing countries.

Depoliticisation and disavowal are further reinforced by the images and text which follow on the final two pages of the report, focused on threats to insurance industries and economies: 'Greenpeace has spoken to people at the top of the

Figure 2.4 Photograph from inside Greenpeace's *Emerging Impacts of Climate Change?*

Source: Greenpeace International 1993. Reproduced with permission.

insurance industry in both London and Zurich who fear for the very future of the insurance industry' (Greenpeace International 1993). Unable to predict the magnitude of catastrophe that climate change may bring, 'the world's biggest business would have been reduced to a shadow of its current self' (ibid). A western economic model is thus presented as a global perspective, where a neoliberal discourse is in tension with a discourse of morality presented through images of vulnerability and suffering. At the end of the report, while Greenpeace identifies human activity as the cause of climate change, it is policy makers who are called upon to take 'real action'.

Discussion – scientific caution and environmental/human catastrophe

The developing visual lexicon of climate change in the early 1990s demonstrates a negotiation by all the actors of the present and the future through imagery. For climate scientists, simulations, in the form of computer modelling and graphs, use present data for future projections. While the images present unseen futures, their validity depends upon observations in the present, constrained by the temporal and visual dimensions of empiricism. Furthermore, their presentation of *global* climate change through simulations of world maps draws upon a history of imperial visions, inscribing power through the observational imperatives of techno-science. Investment in communication and visual technologies as a basis of increasing climate knowledge further invests power in western institutions. The difficulty in understanding these images outside of science is illustrated by the BBC documentary, *The Heat is On* (BBC/OU 1991), which gives credibility to the science of climate by explaining how climate modelling works, and disavowing media representations of extreme weather events in the present. A disconnection between scientific visualisations and media representations is thus established quite early on. However, both science and the media are bound by the use of evidence in the present to make predictions and to help communicate climate change, defined at this stage through science and policy.

Scientific rationality can be contrasted with the rhetoric employed by Greenpeace which seeks to make climate meaningful in the present though the evidential status of photography and an emotional appeal through a language of catastrophe. Although appropriating scientific findings, Greenpeace chooses not to reproduce computer models, instead making the issue meaningful through an array of photographic images that convey the breadth of present climate impacts on humans, animals, habitats and industries, in developed and developing countries. Recent media images of extreme weather events and distant lands are evident here, but this is countered by the wider range of images/effects shown. Yet, the focus upon humans works to reinforce unequal economic and power relations, even as a discourse of morality is present. In all three examples, it is not made clear how the individual and/or communities can help mitigate climate change as the issue is framed through scientific institutions and government policies.

Across all three actors, Manzo's (2010) categories of planet, environment and living beings are represented. The IPCC and the BBC focus upon planet (in the form of globes and maps), and the BBC criticises media coverage of weather events (within the category of environment). In contrast, Greenpeace does not use images of the planet, instead focusing upon the category of environment and living beings to make climate change meaningful in the present. Thus, at this early stage in the developing visual lexicon of climate change, there are distinct differences between these three actors in their choice of images and how climate change is temporally framed. Greenpeace uses photography to lend immediacy, while the IPCC and the BBC adopt a more cautious approach, investing knowledge and power within scientific and computational technology, and the institutions which support them. The purported linearity of technological developments, however, deflects attention away from the urgencies and temporal complexities of ecological time.

Part 2: Scientific caution versus urgent action (1995-1997)

> Detection of a human-induced change in the earth's climate will be an evolutionary and not a revolutionary process…While there is already initial evidence for the existence of an anthropogenic climate signal, it is likely (if model predictions are correct) that this signal will emerge more and more convincingly with time (IPCC 1995, 438).

The characteristically cautious findings of the IPCC meant that its second assessment report in 1995 was only able to state that 'the balance of evidence *suggests* a *discernible* human influence on global climate' (IPCC 1995, 4, my emphasis). The report failed to attract the same kind of media attention as the first had, and was published during a period of low media coverage and public perception. As ever, climate modelling is presented as central to climate science, with much of the report explaining developments since 1990, significantly the production of coupled atmosphere–ocean models and increased confidence in their accuracy as a result of the 'increasing realism of simulations of current and past climate' (IPCC 1995, 5). Although there is still very little regional information available, important progress is identified by the shift from simulating 'global mean changes to comparisons of modelled and observed spatial and temporal patterns of climate change' (ibid). Yet, as 'climate models are calibrated, in part, by introducing adjustments which are empirically determined' (ibid, 34), dependency upon present observations illustrates the temporal limitations of empiricism. While climate models are presented as having significantly improved, they remain constrained by the epistemologies of empiricism where evidence is sought in the present/past. In 1995, the observed data from 1979-1988 is used as a basis for simulations, and as a comparison for assessing their validity, rendering the 'real' as already in the past.

Taking the form of graphs and simulated maps, the images in this report are similar to those produced in the first. What differs is the inclusion of a wider range and more detailed model simulations of the world, with a broader range of colours used to represent predicted global climatic change for different ecological processes, for example, air surface temperature, sea level changes, storm tracks and precipitation. The wider colour spectrum, however, leads to aestheticisation of the images and, potentially, decontextualisation (Hansen and Machin 2008).

The simulated maps of the world depict more detailed outlines of the continents, whilst other variations from the 1990 report include the use of simulated spherical globes, rather than flattened world maps. While regional simulations remain limited, maps of the USA, UK, Europe and Antarctica do appear, revealing the North American and Western European focus of the IPCC. Criticism of the western perspectives presented through the image of the globe as well as the world maps (Sachs 1999; Jasanoff 2004; Cosgrove 1994) thus find further articulation through the regional maps included in the IPCC, reinforced by the dominance of USA and UK scientists as lead and contributory authors to the reports. The IPCC's explicit promotion of the need for more satellite data to provide 'globally consistent data sets of long duration' (IPCC 1995, 528) further links a global perspective to the technological and economic models of the west, inscribing power through visual technologies and perspectives which view the earth from space in order to map and control it. To advance understanding of the climate system the IPCC calls for 'a co-ordinated spectrum of approaches ranging from individual research projects through to global-scale experiments and observational systems and infrastructures' (ibid, 21). The move from the global to the national and local, at this stage, is enacted from the perspectives of the west, communicated through the range of images used in the

IPCC and the dependence upon large scale visual and communications technologies which assign power to governmental and scientific institutions. Although the report calls for substantial cuts in greenhouse gas emissions in order to stabilise greenhouse gas concentrations in the atmosphere, understanding the past, present and future of climate change remains the preserve of science and technology, thereby maintaining the dominance of the west in climate science and in suggested policies.

Although media coverage of climate change was low at the time of the publication of the IPCC's second assessment report in 1995, coverage increased during this period and throughout 1997, as the international Kyoto Protocol was adopted in December 1997 (Carvalho and Burgess 2005; Boykoff and Roberts 2007). The BBC science documentary, *The Science of Climate* (BBC 1997), assigns definitional control of climate change to scientists and scientific technologies. Opening with an image of a satellite circling the globe, and news images of floods, hurricanes and storms, the narrator tells the viewer that satellites collect weather data 24 hours, but that climate is a complicated system. Initially making climate culturally meaningful through weather, the premise of the programme is to complicate this interpretation through a consideration of the complexity of climate beyond weather. It does so by foregrounding the scientific technologies and practices required to understand climate, and the necessity of acquiring empirical data. Dominating its visual narrative are images of satellites stationary in space and orbiting the globe, computer models in the form of globes, and graphs depicting temperature changes. These scientific representations position scientific technologies and practices as in control of climate, as the documentary seeks to downplay the urgency of climate change through an appeal for more data and interdisciplinary research, so that scientists can better ascertain the extent to which climate change is the result of human causes. The scientific representations are supported by footage of male scientists working in situ, for example, looking at ice cores in Antarctica, or men working on computers. While there is some discussion of the potential human causes of climate change – 'mankind is burning fossil fuels' – overall, discourses of science and technology prevail, as the urgencies of addressing climate change are over-ridden by the faith in the rationality of the practices and protagonists of science and technology. In doing so, the viewer is left unable to contribute to the mitigation of climate change, as confidence is invested in the retributive qualities of science.

Investment and faith in the techno-practices of (climate) science can be contrasted with the approaches taken by Greenpeace in 1997, through its international campaign to stop new oil exploration. *Putting the Lid on Fossil Fuels* (Greenpeace UK 1997) renders climate change present by linking fossil fuel production to climate change (see Figure 2.5). Although the climate models on the front cover appropriate scientific authority, they function as catalysts for immediate action, not for more investment in climate modelling. Greenpeace utilises the representational practices of science in order to give legitimacy to its efforts to make climate change meaningful. The image of the globe draws more explicitly upon discourses of global interconnectedness and planetary vulnerability, common to NASA's *whole earth* image (Cosgrove 1994), but reappropriates these meanings in the context of climate change. The nearly uniform redness encircling and shrouding the once-beautiful planet of blue, green and white in NASA's original image signifies a globally uniform increase in temperatures that appears horrifying in its magnitude. Drawing upon the meanings associated with an existing photographic icon central to the discourse of environmentalism, Greenpeace gives meaning to climate change in the form of danger and catastrophe (similar to its earlier communication strategies). Even though the photographic image is digitally altered, and is not 'real', it functions both denotively and connotively through its referential relationship to the original photograph.

Figure 2.5 Front cover of Greenpeace's *Putting the Lid on Fossil Fuels*
Source: Greenpeace UK 1997. Reproduced with permission.

The reappropriated planetary icon gives meaning to, and is given meaning by, the set of four globes appearing underneath it. Representative of climate modelling, these globes are more explicitly linked to scientific discourse, using the conventions of scientific illustrations/graphs through temperature data and time. They lend (scientific) legitimacy to the enhanced photographic globe while, simultaneously, the image of a globally warmed world gives meaning to the computer models through its use of environmental discourse. Yet, 'the significance of the globe in the language of contemporary debate about the environment' is problematic, because it renders the world 'as an object of contemplation detached from the domain of lived experience' (Ingold 1993, 31-32).

Climate science is given an authoritative voice throughout the report, with quotes from the IPCC's second assessment report on the opening page: '"The balance of evidence suggests a discernible human influence on global climate" IPCC, 1995' (Greenpeace UK 1997). This science is somewhat simplified by Greenpeace's translation of this as affirmation by the IPCC that 'climate change had arrived' (ibid). In contrast to the truth status assigned to climate change scientists, Greenpeace critiques the politics of climate change by appealing to the world's governments to 'take these opportunities to agree a fossil fuels budget for the world and commit to a phaseout' at the forthcoming Kyoto Climate Convention in December 1997 (ibid).

The report offers a linear visual narrative of cause, effects and solutions. The inclusion of photographs of oil platforms on the opening page is intended to establish an associative link between oil and climate change. The effects of climate change are conveyed by a series of four photographs placed together, depicting devastation wrought to buildings and landscape by wind and water; indexical signs of an already warming world. At the end of the report a photograph of a group of children surrounding a solar panel is used to metonymically signify the future, and solution to climate change. The temporal dilemma which underpins climate change communication, that is, the need to communicate both the present reality and future threats, without extensive visual evidence, is negotiated in the report through the inclusion of photographs of existing destruction wreaked by weather, alongside the future possibility of a warming planet signified by the computer model predictions.

The tying of the impacts of climate change to a particular location – the Atlantic Ocean – utilises the more usual discursive strategy of aestheticising nature and humanising animals (Soper 1995) to create an emotional resonancy. The beauty and pristineness of a specific ocean habitat is set in visual opposition to industrial oil pollution, 'a wilderness which the industrial world can well afford to leave alone' (Greenpeace UK 1997). The burning of fossil fuels as the cause of climate change is communicated through the destructive image of an oil platform producing a fiery plume, with the risk of flood and hurricane damage communicated via photographs. Significantly, humans are not identified as victims, rather the loss of beautiful ocean habitats and its wildlife, with the clean technology of solar panels on the final page setting up an opposition between dirty and clean energy.

Discussion – mapping climate change through simulated globes and animal habitats

During the mid 1990s, climate modelled globes produced by climate scientists and reproduced in the IPCC report move from the domain of science into televisual narratives and NGO communication on climate change, acquiring a range of different meanings according to their discursive context. In the IPCC report, the simulations gain legitimacy as representations based upon observed, empirical data, but are simultaneously constrained by reliance upon observed data in the present/past. Furthermore, these visualisations reinforce a particular view of the world, and of climate change, as able to be understood, mapped and (ultimately) controlled through expansion of computing and satellite technology. Although attempts are made to offer a more spatially nuanced range of predictions through regionalisation, the focus upon the UK, USA and Northern Europe reproduces a western perspective. Televisual narratives bring the science of climate and its visualisations to a mainstream audience, yet the cautious findings of the IPCC are reproduced through the BBC documentary analysed, offering a conservative and science bound view of climate change.

The reproduction of climate models in Greenpeace's campaign material signals a more radical reappropriation of their meanings. While the IPCC report and BBC documentary take time to explain the complexity of these models in relation to the accumulated data they are based upon, Greenpeace positions these models as evidence of global climate change in the present as a catalyst for immediate action. In drawing upon the meanings associated with the existing icon of the planet, the globally warmed planet draws upon the viewers' imagination in a way that makes climate change more real through its link to a national (UK) context. Greenpeace's continued use of photographic images reinforces its investment in making climate change real in the present.

To some extent, this period marks a shift towards linking climatic change to particular regions and places, through the start of regional climate modelling, and Greenpeace's focus upon a specific geographical area close to the UK. However, the focus upon animals and their habitats more specifically, marks a slight shift away from the impacts upon humans present in its previous campaign material. As the IPCC and BBC continue to image climate change through computer modelling in the form of globes and maps (at the expense of environment and living beings), Greenpeace also simplify their visual language to focus upon ocean animals. While science and the media continue to link science to policy, environmentalists politicise climate change by identifying the links between industry and government in the production of greenhouse gases.

Part 3: Visible 'evidence' in the present – distant ice, local weather (1997-2001)

Linked to the 1992 United Nations Framework Convention on Climate Change, the Kyoto Protocol was established in 1997 in order to set 'binding targets for 37 industrialized countries and the European community for reducing greenhouse gas (GHG) emissions…at an average of five per cent against 1990 levels over the five-year period 2008-2012' (UNFCCC 2009). Although the protocol did not enter into force until 2005, and was not adopted by Australia (until 2007) or the USA, its establishment signalled an important international governmental commitment to dealing with climate change through the reduction of greenhouse gases. Although receiving high media coverage in the lead up, following the announcement of the protocol there was a drop in media coverage of climate change until 1999, when there was a dramatic increase until 2001, the year the IPCC's third assessment report was published (Carvalho and Burgess 2005; Boykoff and Roberts 2007). Yet, this period of fluctuating media coverage marks an important shift in the visual discourse of climate change towards identifying more geographically specific places as at risk from climate change in efforts to make it more personally meaningful.

For Greenpeace, during the latter years of the 1990s and into the 2000s, the Arctic and Antarctica figure as geographically and politically contested, sites within the mitigation and communication of climate change. The Arctic, and Alaska in particular, is not only an area at the centre of oil expansion, but also an area that has experienced the effects of climate change more acutely than others. As oil companies such as ARCO, BP, Shell and ExxonMobil are targeted, the Arctic and its inhabitants also begin to figure prominently in the iconography of climate change. In July 1997, Greenpeace set off on its first expedition to Alaska's North Slope in the Arctic, 'to study the impacts of global warming and to confront offshore oil development' (Greenpeace International 1999). Both scientific expedition and environmental action, subsequent trips took place over the next three years, specifically to prevent BP from constructing its Northstar oil field project. Both humans and animals are represented as at risk from the impacts of climate change: the lands and cultures of indigenous Alaskans, and animals such as the polar bear and caribou. While the impacts upon the inhabitants of Alaska are being experienced and felt, in Antarctica more potent visual 'evidence' of climate change emerges. The crack in the Larsen B ice shelf, identified and photographed by Greenpeace in 1997, signifies as visible evidence of increasing temperatures, thus initiating one of the most dominant sets of images of climate change, that is, melting/retreating ice, which effectively supplants the human voices of climate change.

Nature's Bottom Line: Climate Protection and the Carbon Logic (Greenpeace UK 1998) makes full use of the photograph of the crack in the Larsen B ice shelf (see Figure 2.6). It denotes the impacts of climate change and, drawing upon dominant discourses of the environment, connotes a fragile and visible nature at risk. Stretching from the bottom of the page across to the top, it draws the eye beyond the physical borders of the image, making the damage to the ice appear extensive. Although the inside of the report explains in detail how much carbon, in the form of fossil fuels, we can afford to burn in order to stay within 'nature's ecological limits', the photographic image implies that we are already beyond those limits. From the heat and warmth explicitly connoted by the cover of its previous publication (*Putting the Lid...*), where climate change is implied as a future present, the temporal discourse shifts to impacts in the present. However, the crack in the ice signifies as evidence of increased temperatures, and global warming/climate change, only in relationship to previous associations with red heat.

Figure 2.6 Front cover of Greenpeace's *Nature's Bottom Line: Climate Protection and the Carbon Logic* showing crack in Larsen B ice shelf

Source: Greenpeace UK 1998. Reproduced with permission.

Although the documentation of visible impacts enables Greenpeace to shift attention towards promoting solutions through renewable energy, it is interesting to note how this period in its campaigning (from 1997-2000) has appeared to leave a lasting impression on the wider visual lexicon of climate change, that is, the production of the Arctic/Antarctica as a 'symbolic/iconic' environment of climate change (Hansen and Machin 2008, 792). Although Greenpeace focused upon human and animal impacts, its imagery links the Arctic to the polar bear more specifically. Since achieving iconic status as the 'poster child' of climate change, the polar bear has diverted attention away from the impacts of climate change upon humans, particularly the cultures and livelihoods of the indigenous peoples of the Arctic and sub-Arctic (Trainor, S.F. et al. 2007).

The shift towards more geographically specific impacts of climate change in the visual discourse of Greenpeace's climate change communication is also adopted by BBC documentaries. Rather than distant (to UK citizens) lands, peoples and animals, however, the impacts of climate change in the UK are foregrounded. In *Underwater Britain* (BBC 2000), climate change is linked to specific weather events, previously criticised as media misrepresentations by the IPCC and earlier BBC documentaries. Opening with the observation, 'more storms, more floods… It looks like the first instalment of global warming' (ibid), the documentary implies that these may be the first impacts of climate change in the UK, using weather events as evidence of climate change; an association that has a longer history in news media and NGO representation. Significantly, this is a human drama, following the lives of people affected by widespread flooding, particularly in Selby, Yorkshire, UK. Here, a warlike mentality is connoted through footage of the army bringing in sandbags to stem the floods, with families helping to fill the sandbags in communal efforts reminiscent of the Second World War. Framed through a discourse of war, the map of Britain is invoked as a territory that will be geographically and economically reshaped as a result of a changing climate. The narrator states that insurance companies are already redrawing the map of Britain, as more areas are identified as at risk from flooding, while the physical contours of Britain, specifically East Anglia, will have to be redrawn as low lying land is submerged under water. The message is that climate change impacts are happening and that this is a present indication of Britain's future.

The human experiences of flooding – as individuals and families are interviewed in their flooded homes, or returning to them once the waters have receded – gives an emotional resonancy to the issue of climate change, supported by the rationality of science. When one resident says, 'Someone needs to tell us if it [flooding] is global warming', the shot cuts to Professor Martin Parry, a climate scientist and a contributor to the IPCC's fourth assessment report (2007), who declares, 'it fits the pattern' (BBC 2000). Computer models from the British Met Office are then shown, as a representative, Helen Young, tells us that there will be more and heavier rain. The narrator states that 'the scientific community now largely agree that global warming is here' (ibid). Scientific discourse and representational practices are offered as rational and authoritative, and computer model simulations are able to signify as more real and relevant in relation to local experiences of climate change.

Like scientists, politicians are given positions of authority in the programme. A flood victim is shown in conversation with Deputy Prime Minister, John Prescott. Prescott asserts that the flooding is due to climate change, in contrast to one resident who says they do not believe him. Scientists and politicians are presented as the rational voice of climate science, while the public are figured as emotional sceptics. The programme thus sets out to make the case for the reality of climate change using the emotional resonancy (and partial scepticism) of the public for the viewer to potentially identify with, in contrast to the assertions of scientists and politicians. Significantly, institutionalised voices from science and policy frame the issue, while NGO voices are not represented.

The documentary leaves the viewer unclear as to what they can do to help mitigate climate change. Instead they are left with the notion that they may be at risk and are vulnerable, as '1 in 10 people in Britain live on floodplains' (ibid). More commonly used to represent people in developing rather than developed countries, this sense of vulnerability is reinforced through the temporality of climate change. While the programme reinforces the reality of climate change in the present, the certainties of the past are said to be giving way to the uncertainties of the future. Climate change is framed as unpredictable, unprecedented and unknown, with worse to come. The viewer is left with a sense of foreboding and dread that does little to help with the idea of mitigation.

Discussion – distant polar bears and localised human drama

During the latter years of the 1990s there emerges a distinct visual lexicon of climate change that continues its resonancy today. The symbolic and iconic environment of the Arctic/Antarctica figures prominently as a geographical and imaginative place at risk from increasing temperatures, with the polar bear as the emerging icon of climate change. Climate imagery becomes associated with distant places and loss of animal habitats. At the same time, attempts are made to make climate change meaningful and relevant by focusing upon national/local contexts. In the case of the BBC, the extreme weather events of flooding are presented as the human drama of climate change, supported by scientific discourse. The status of the science of climate change is thus more or less established as a consensus in mainstream representation; a position that Greenpeace had taken since the early 1990s.

Using Manzo's (2010) typology, the categories of planet, environment and living beings are all utilised, but represented in very specific ways. Representing the category of planet, the globes and maps of science continue to be reproduced by BBC documentaries but are given meaning through their specific relation to the UK in the form of regional climate modelling. Maps are literally and metaphorically redrawn; through the changing physical contours of the UK as a result of rising sea levels, and the economic redrawing of areas at risk in the UK through increasing flood. The environment and living beings feature in specific relation to each other. Thus, the Arctic as a geographical space is visually and imaginatively represented by polar bears, effectively shifting attention away from humans. Simultaneously, the move towards specific landscapes in the UK, and the impacts to humans through extreme weather events, finds representation. The BBC's criticism of the media dramatisation of extreme weather events as representative of climate change is reversed, as local human drama is offered. However, the BBC documentary requires the evidence provided by scientists to be able to present these events as signs of climate change, continuing its discursive framing of climate change through science.

This period marks a shift towards imagery linking the global to the local. Yet, as O'Neill and Nicholson-Cole have shown through focus group discussions in the UK, the juxtaposition of images of distant lands/animals and human suffering at a local level, although making climate change seem more important, also disempowers people, making them feeling unable to address the issue (O'Neill and Nicholson-Cole 2009). Such disempowerment is also indicative of how responsibility for mitigating climate change is discursively framed. While Greenpeace identify industry and politics as the basis for change, the BBC continues to provide no critique of industry or governments, nor offer a means for the viewer to help mitigate climate change. Instead they are presented as (sceptical) victims of a decontextualised and depoliticised issue.

Part 4: Observing climate impacts in the present/past (2001-2007)

> An increasing body of *observations* gives a *collective picture* of a warming world and other changes in the climate system (IPCC 2001a, 2, my emphasis)

In 2001, the IPCC's third assessment report finally declared that 'there is new and stronger evidence that most of the observed warming observed over the last 50 years is attributable to human activities' (IPCC 2001a, Preface ix). The period following the publication of the report, and leading up to its fourth assessment report in 2007, marked a distinctive shift in political and media discourses on climate change. A significant increase in print media coverage in Western Europe and North America occurred, particularly so after 2003, following summer heat waves in Europe and Hurricane Katrina in the USA in 2005 (Boykoff and Roberts 2007). Although in opposition to the cautious findings of the IPCC, and supported by the discourses of earlier BBC documentaries, the linking of extreme weather events to climate change gained momentum in the news media, as climate change impacts became increasingly linked to geographical places other than the Arctic/Antarctica. Political debate about the urgency of mitigating climate change was aided by the publication of the UK Government's Stern Review (2006), which provided an economic incentive for addressing climate change, while the release of Al Gore's documentary, *An Inconvenient Truth*, in September 2006, offered further

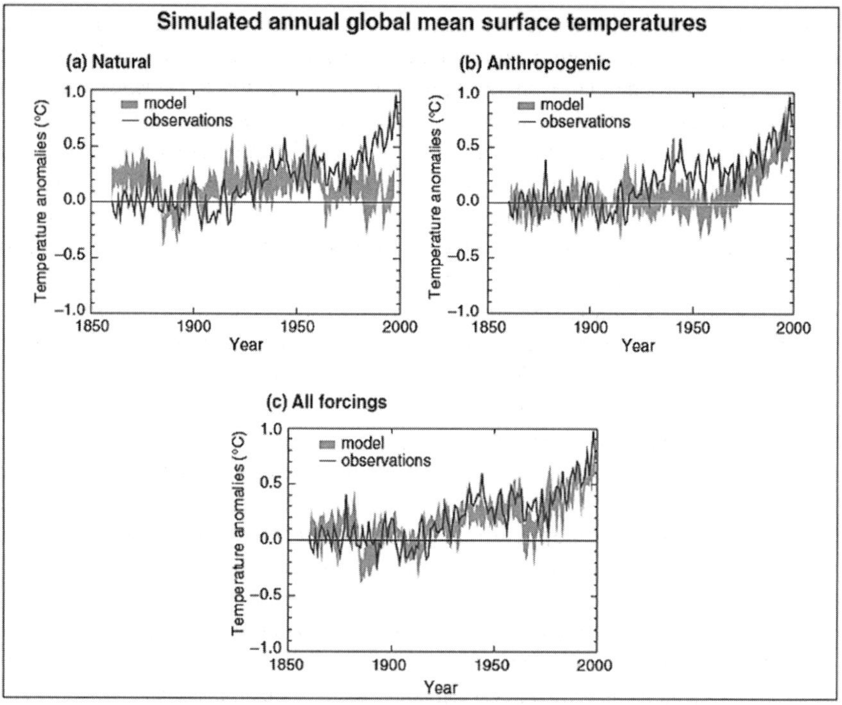

Figure 2.7 IPCC graphs showing visible correlations between observed and simulated temperature rises from 1860-2000

Source: *Climate Change 2001: Synthesis Report. A Contribution of Working Groups I, II and III to the Third Assessment Report of the Intergovernmental Panel on Climate Change*, Figure SPM-2. Cambridge University Press. Reproduced with permission of the IPCC.

high profile representations of climate change. Yet, political debate continued with regards to mitigatory action, with the UK Government attempting a controversial rebranding of nuclear power as carbon neutral (Bickerstaff et al. 2006; Doyle 2011). Therefore, while the significance of climate change increased in terms of media coverage, political debate and public understanding, the different discursive framings of climate change continued.

What had made the IPCC finally attribute climate change to human activity? This was related to 'new and stronger evidence' (IPCC 2001a, Preface ix), provided by correlations between observed and simulated data of climate processes (see Figure 2.7). As in the previous assessment reports, graphs and computer modelled maps feature prominently, but in this instance these representations are presented as more 'real' than in previous reports. What are the conditions of this realism, as the new evidence of human induced climate change? Since the 1990 first assessment report, observational data had been used to simulate future climatic changes, with the persistent message of the IPCC reports being a call for investment in the collation of more data and better computational technology. In 2001, the claims that increasing temperatures were the result of increased greenhouse gases in the atmosphere through human activity are based upon the results of climate modelling matching the observed data. In other words, the direct observations match the simulated models because, by 2001, there is more empirical data to draw upon and computer modelling is more technologically advanced. As such, the conditions of realism are empirical data and computational ability, as science and technology are intertwined as discourses of truth.

Temporal disjunctures are again revealed as the representational limitations of climate science. Investment in technology is an investment in developments over time, while the need for more empirical data on climate processes also takes time for these to emerge and to be gathered. The point at which enough observational data is accumulated, and computer technology is sufficiently advanced to be able to process this data, for scientists to be able to declare that climate change is attributed to human activities, is at temporal odds with the progression of climate change. The IPCC state, '*several centuries after* CO2 emissions occur, about a quarter of the increase in CO2 concentrations caused by these emissions is still present in the atmosphere' (2001a, 17, my emphasis). Furthermore, 'ice sheets will continue to react to climate warming and contribute to sea level rises for *thousands of years after* climate has been stabilised' (ibid, my emphasis). Temporal disjunctures between climate processes and scientific knowledge production reveals the problem at the centre of scientific approaches to understanding ecological issues. In the time taken for the IPCC to attribute climate change to humans, greenhouse gas emissions continued to rise, further committing the world to future climatic change.

As a science-policy document, the IPCC produced a synthesis report of the findings of its three working groups (IPCC 2001b), increasing its potential mediation across science, policy, media and public domains. As a form of science communication, the synthesis report is noticeable for its aesthetic and stylistic decisions, as coloured borders and typeface draw attention to specific findings. Increased aestheticisation is also apparent in the wider range of colours used in the graphs and computer models, as the communication of climate science involves representational decisions that take into account the visual format of the findings. Furthermore, the production of knowledge is itself guided by choices of representation, where 'the need to balance scientific understanding against computational efficiency and model realism often guides the choice of the particular class of models used' (IPCC 2001a, 475). In 2001, the deficiencies of computer models, however, still relate to the question of scale, as regional models remain difficult to produce.

The political conservatism preventing the IPCC from attributing climate change to human activities until 2001 has been reflected in the BBC documentaries on climate change since the early 1990s. Yet, the titles of a high profile two-part BBC (in conjunction with Discovery Channel and Open University) documentary series,

Are We Changing Planet Earth? (BBC 2006a) and *Can We Save Planet Earth?* (BBC 2006b), frame climate change as an issue still open to scientific debate. Written and narrated by natural historian Sir David Attenborough, they are both science and natural history documentaries, utilising an expository mode by putting 'the case for a particular view or interpretation of evidence before us' (Nichols 2001, 4). It is Attenborough's own journey as a natural historian to accepting the reality of climate change (a narrative device also used in Gore's *An Inconvenient Truth*) that provides the basis for evidence and interpretation. Supported by an explanation of climate science, video evidence of climate change impacts and the presentation of potential solutions, Attenborough's exposition draws the viewer along his journey to accepting the 'truth' of climate change. The organising logic is both a story and the solving of a problem (Nichols 2001). As a veteran producer of BBC natural history documentaries over 50 years, Attenborough's professional and cultural status provides a voice of authority, reinforced by the scientific discourse Attenborough draws upon and embodies. The truth status of the figure of Attenborough and science is substantiated further in the commercial DVD set of the programmes, packaged and sold under the title, *The Truth About Climate Change* (2008).

The textual rhetoric is supported by the visual rhetoric, which reproduces all of the categories of climate change imagery identified by Manzo (2010) – planet, environment and living beings – and utilises a range of visual technologies and practices including video footage, satellite imagery, computer simulations and scientific graphs. Yet, the most dominant sets of images used are video footage of landscapes and animals, alongside computer model simulations of a warming planet and scientific graphs of global temperature increases. The framing of climate change is one in which visible (video) evidence of climate impacts upon animal species and their physical habitats is prioritised, alongside the truth claims of scientific visualisations and data in the form of computer model simulations and graphs. The 'subjunctive' (Wolf 1999) nature of the computer simulations draws upon the indexicality of video footage, and vice versa, to frame these images as 'truthful' representations of past, present and future climate change. Through the images and discourses of science and non-human nature (environment), the link between representation and reality is established, further reinforced by the authoritative status of Attenborough as spokesperson for the natural world, and through 'extensive footage of interviewees presented as experts', that is, scientists (Mellor 2009, 143).

Temporal and visual discourses intertwine, as past, present and future are represented through visual practices. The first documentary draws upon visual and empirical evidence in the present and past to answer the question posed by the title – *Are We Changing Planet Earth?* (BBC 2006a) – in the affirmative. Initially this is undertaken via Attenborough's reflections upon his own history of documentary film-making, particularly his exploration of the developments of life on earth before humans. Focused primarily upon animal and plant species, Attenborough states in the opening sequence: 'Since life began four thousand million years ago, it [planet earth] has gone through extraordinary changes in relation to climate and the species that live on it' (ibid). Ecological time is thus established through clock time, where the 'the arrival of human beings' is 'still a recent event in the history of our planet' (ibid). Although changes to the climate and species have always occurred, the fairly recent and rapid changes are attributed to the activities of humans: 'it seems that our planet is being transformed, not by natural elements, but by the actions of one species, mankind' (ibid). As a species, humans are presented as newcomers to planet earth who have brought potentially devastating effects.

As a dominant discourse of the environment, the imposition of humans on the planet is simultaneously reinforced and undermined by the visual rhetoric accompanying the narration. Opening with an aerial viewpoint from outer space, the camera looks down onto the edge of planet earth, then zooms into this earth, traversing above and along a series of generic landscapes that represent the visible beauty of the natural world. Accompanied by reverential music, Attenborough slowly states, 'This is our planet – planet earth' (ibid). Visualising the beauty of nature through landscape and animals, the narrative invites the viewer to consider the earth as 'ours', suggesting human ownership of the natural (non-human) world, even as humans are identified as a recent and damaging addition. Introducing the history of animal and plant species, the visuals cut to polar bears on ice. In stark contrast to the visible beauty of such animals, humans are represented by images of urban cities. The opening sequence thus reproduces dominant discourses of the environment through the icon of planet earth, to simultaneously connote the beauty and fragility of this earth through the landscapes and animal/plant species that inhabit it, against the potential destructiveness of human beings. As such, the viewer is addressed as both an observer/owner of this planet and its beautiful species, as well as a potential destroyer. 'We' are simultaneously part of this planet, but also outside, looking down on its visible beauty, separated and disembodied from nature.

In setting out to answer the programme's question, Attenborough establishes a strong visual aesthetic that reproduces dominant discourses of the environment as a visible nature at risk established in the opening sequence. In the first 20 minutes the viewer is offered a visual tour of the world's animal and plant species, alongside footage of extreme weather events, as Attenborough states, 'In every part of the world new climatic extremes are now being recorded. Our weather is in turmoil' (ibid). The indexical nature of these images function as mounting evidence of potential climatic change, as polar bears are identified as the first victims of increasing global temperatures. The visual and narrative prioritisation of this species in the programme's rhetorical structure, followed by the retreat of mountain glaciers, renders climate change, for the viewer, as distant and 'out there'. Indeed, while human voices are accessed, through interviews with the inhabitants of the South Pacific island of Tuvalu, whose homes are routinely flooded as a result of rising sea levels, and through video footage of the impacts of Hurricane Katrina on New Orleans (USA) in 2005, the visuals, reinforced by the opening sequence of planet earth, assign climate change to dramatic weather events.

Although the documentary offers these images and events as sets of visual evidence, Attenborough continues to question their validity, drawing the assumed viewer along with him on his personal journey: 'is global warming and the changes we are now seeing really a new phenomenon in the earth's history?' (ibid). Further evidence is sought and acquired through scientific data, supported by interviews with a range of male scientists. Empirical data is presented in the form of atmospheric measurements and glacial ice cores, supported by technology in the form of super computers that 'can perform 10 billion calculations per second' (ibid). Discursive practices of techno-science reinforce the visible evidence provided by video footage of weather events and impacts to animals, plants and, to a lesser extent, humans, in order to build an evidential case for the reality of human induced climate change.

The interconnections between time and vision are a significant discursive feature. Empirical evidence is sought in the past and the present to provide the case for present and future climatic change. Attenborough looks back at his own documentaries in order to show the changes over time, now visible in the present. Archive footage of Attenborough's *Life on Earth* (1979) series, where he explored the beauty of coral reefs, is compared to the bleaching of corals in the present as a result of increased ocean temperatures. Similarly, footage of Attenborough in the treetops of the Amazonian rainforest 27 years earlier is contrasted with the current state of the rainforest as a result of deforestation and drought. Signifying as before and after shots, present conditions are authenticated through reference to the past based upon differences that have become visible over time. Yet the visualisation of the passing of time, charted through observable change, is problematic in the context of climate change. Incorrectly, one of the scientists states, 'What we are *seeing now* is the impact of humans, humans are *beginning* to change the climate' (BBC 2006a, my emphasis). What we see in the present, however, are the effects of past emissions (Kolbert 2009), as humans began to change the climate before these changes could be seen. Reinforcing the dominant discourses of science as empirical data observable in the present/past, the documentary fails to acknowledge the invisibilities and latencies which characterise climate change.

Both a knowledge structure and representational practice, scientific clock time is deployed to understand and validate climate change through the deployment of graphs. While the first documentary began with Attenborough offering a visual and textual synopsis of the development of animal and plant species over time, the programme ends with a graph depicting the increase of global temperatures over time, from 1850 to 2000. In an effort to make the scientific graph more meaningful, both Attenborough and climate scientist, Professor Peter Cox, literally insert themselves into the graph marked out on the floor of a hall. Cox says 'what we are going to do is take a walk through time, and the first thing to note as we walk through is that climate is naturally variable' (ibid). Walking from left to right to demonstrate the passing of time, a growing red line is superimposed onto the floor to depict the rate and levels of temperature rise. When the line reaches 2000, the graph depicts an overall increase in global temperatures. Two further coloured lines are then superimposed upon the graph: green represents the data from a climate model which takes into account natural climate variations; yellow represents the data from a climate model which takes into account natural variability and human factors such as the greenhouse gas effect. The yellow line visually resembles the trajectory of the red line, depicting human attribution of increased global temperatures since 1850. As a simulated representation, the yellow line on the graph is given the status of indexicality when Cox states, 'the model looks realistic, it looks like the real world' (ibid). Given authority by science, this slippage between representation and reality is afforded further credence by Attenborough when he declares, 'So there you have it. There seems little doubt that this recent rise, this steep rise in temperature, is due to human activity' (ibid). The first documentary thus uses a combination of scientific visualisations and practices in the form of graphs and the gathering of quantitative empirical data, as well as old and new video footage of animals, plants and extreme weather events, to answer the question. The temporal discourses of past and present are communicated through imagery to validate human induced climate change

In an effort to identify solutions to climate change, the second documentary – *Can We Save Planet Earth?* (BBC 2006b) – switches the temporal dimension to the present and future. Clock time underpins the textual and visual rhetoric. The opening sequence reiterates how changes in the past have been natural, while current changes are caused by humans. What have now become dominant images of climate change are deployed in the opening sequence: penguins, polar bears and cubs, satellite images of planet earth, bleached coral, waves, storms, and people stranded in floods. Attenborough's accompanying narrative interchanges between the temporal dimensions of present and future: 'We *are* causing the world to heat up', 'The vast forest of the Amazon *could* wither and burn', 'The oceans *could* turn acid', 'The arctic *could* be transformed', 'Rising tides *could* cast millions of people adrift', 'There *are* ways to minimise these changes' (ibid, my emphasis). The move from present to future implies the possibility of positive change; Attenborough says, 'We still have time' (ibid). Potential and imminent catastrophe is down played by the possession of time.

Attenborough reiterates the scientific consensus on climate change introduced in the first documentary, but simultaneously undermines it by partially removing human responsibility: 'The great majority of scientists are now agreed that our planet is warming and that human beings *are at least some part of the reason* it's doing so' (ibid, my emphasis). In contrast to the first, however, this documentary focuses more specifically upon humans as contributors to, and victims of, climate change, although the overall emphasis is upon contribution rather than impacts. Throughout, video footage of existing impacts upon humans, such as the flooding in Boscastle, UK (2004), Hurricane Katrina, USA (2005) and increasing bush fires in Australia, are juxtaposed with visualisations of the future in the form of computer model simulations of a warming planet and mock British weather forecasts for the year 2050 of high temperatures and increasing storms. The continuous movement between timescales of past, present and future, and between different geographical places, is a convention of expository documentary 'which may sacrifice spatial and temporal continuity to rope in images from far-flung places if they help advance the argument' (Nichols 2001, 107). The argument that climate change is real is articulated through the evidentiary status of video and simulated images.

Visualisations of time (as present and future) take centre stage, as seeing and sight are central to the legitimation of climate change. Computer simulations make visible the invisibility of greenhouse gases, by showing a grey blanket around the earth to represent the thickening of the atmosphere, and through simulated black blocks to signify emissions spewing from households, cars and factories. Future climate change is visualised most prominently by climate models which constitute a large portion of the visual and textual narrative of the documentary. Invited by climate scientist Cox to 'help me visualise what's in store for us' (BBC 2006b), Attenborough and Cox are shown sitting in front of a screen at a projected climate model. This image of the two scientists sitting in a darkened room contemplating the large model projection is continually returned to at points throughout the programme, as the conversation between the two interjects the exposition. The sheer size of the projected image, and the continual return to this scene, gives the image a sense of importance, as it gains its authority through discourses of science and visual evidence, in order to give credibility to future predictions. Attenborough claims, 'This is a vision of how and where the world will heat up if we continue down the path we are on' (ibid). Yet, at the same time, this constant *looking* at the simulated image of the map of the earth as a means of understanding how the world will warm is both a disembodied and decontextualised view, similar to looking at the icon of the planet earth; we are positioned as viewers onto a world of which we are simultaneously part of, but disconnected from. This is reinforced further through the assumption that viewing a world map, where rising temperatures are indicated by changing colours across continents, can tell us about what it will be like to live in a world of increased temperatures, decontextualised from the lived, everyday experiences of a climatically changing world.

Changing individual behaviours is the main way that humans can address climate change, specifically the carbon emissions produced by 'domestic activities'. Through the story of a fictitious western family – The Carbons – the viewer is taken on a visual journey of domestic carbon production. Comprised of a heterosexual couple with two female children, living 'in an average suburban house outside an average city' (the house is by no means of average size), and using 'an average amount of electric power', they signify as a 'normal' family based upon traditional gender roles (ibid). Mother is shown shopping, as Father travels to work by car and takes a business trip by flight. Constructing an assumed viewer through a heterosexual moral economy of the household, their normativity goes unquestioned. Car use, flying, electric appliances such as kettle, oven and TV, packaged food flown across large distances and food wastage are identified as carbon producing practices. These everyday practices of energy consumption through domestic appliances are linked visually and textually to the melting glaciers of the Arctic and Antarctica, and to drought in the Amazon rainforest, in an attempt to link the local with the global, the immediate to the distant. Yet, in promoting a discourse of individual lifestyle for the mitigation of climate change, responsibility is placed on the individual, at the expense of wider structural and institutional changes such as government policy, or business practices (Hobson 2006).

Accompanying this discourse of individualism is the concomitant message that not much behavioural change is required. Attenborough is shown around a typical UK house by another expert, Dr. Dave Reay from University of Edinburgh, to illustrate what changes can be made to reduce carbon dioxide emissions, including turning off lights and appliances when not in use, turning down the thermostat, switching to energy saving light bulbs, composting waste, buying an energy efficient fridge, or only boiling the specific amount of water needed in a kettle. These changes represent 'simple and painless steps', often encouraged by Government climate communication campaigns, for example, the UK Government's 'Act on CO2' (UK Government 2009) and the European Commission's 'You Control Climate Change' (European Commission 2009). Such approaches are criticised by WWF as inadequate because of their failure to facilitate deeper, more 'ambitious' behaviour change, including an 'engagement with political process' (WWF-UK 2009, 6). Yet, talking to Reay, Attenborough says, 'Alright, let's get this clear. It's not about stopping ourselves making cups of tea, we're talking about *living the life more or less as we've lived it before* but more efficiently, and waste is the source of our trouble, not actually living?' (BBC 2006b, my emphasis). Answered in the affirmative, imminent catastrophe is averted through an appeal for individuals to make small and painless behavioural changes to their everyday lives.

The profound disconnect of people's lives from the reality of climate change is further reinforced by the foregrounding of technological salvation, from small to large scale. Much of the second documentary is concerned with identifying solutions to climate change, as scientist, Professor Steve Picala, presents 7 steps to stabilising greenhouse gases on current levels over the next 50 years (a figure that is in opposition to the stabilisation of greenhouse gases to 1990 levels that is needed and proposed by the IPCC and the Kyoto Protocol). Quantifying the reduction of greenhouse gases into identifiable slices on a graph of rising temperatures, each solution proposed enables one slice to be removed, presenting a linear process of cause and effect in the mitigation of climate change. Behavioural and technological changes are proposed, from small scale behaviours to large scale technologies. Changing domestic appliances and behaviours in the house remove ones of the imaginary slices, using a fuel efficient car removes another, but is dependent upon '2 billion of the world's cars' being able to do '60 miles to the gallon' by 2050, thus placing trust in future technology (ibid). Driving less removes another slice, but contradicts the investment in car manufacturing previously proposed.

Discourses of technology prevail, as large scale technologies are presented as offering the biggest means of emissions reduction. Under the guise of 'carbon free sources of power' such as solar and wind, nuclear is proposed; 'however controversial, tripling the world's nuclear power would cut our emissions by a further slice' (ibid). The discursive construction of nuclear as carbon free is a fairly recent development, promoted by the UK Labour Government as a means of building public support for the commissioning of new power stations; and representing a U-turn on their commitment to no new nuclear power (Bickerstaff et al 2006; Doyle 2011). The documentary thus reproduces dominant governmental and institutional discourses without acknowledging the carbon produced through the building of new power stations (which can take up to ten years), nor the significant amount of financial investment required to build new nuclear stations, at the expense of investment in renewable technologies such as wind and solar. Techno-fixes are further promoted through the unproven technology of carbon capture and storage (CCS), introduced by Attenborough as 'technology we already have [that] can save us' (BBC 2006b). Picala states 'the Norwegians have estimated that we could store all of Europe's carbon dioxide for the next few hundred years in this way', describing it as 'exactly the kind of technology that could keep those monsters from ever coming out behind their doors', monsters being 'things we don't yet fully understand' (ibid). The inaccurate presentation of CCS as proven technology is further obfuscated by the reference to unidentified monsters, which the technology will help keep at bay. As this is the last technology presented in the discussion of solutions to climate change, it serves to undermine the behavioural changes which preceded it, and the viewer is left with the incorrect impression that technology will store the carbon we emit.

The two documentaries are reliant upon video evidence and scientific simulations to make climate change 'real'. Through an expository mode, in the context of science and subjunctive documentaries, the viewer is taken on a journey by Attenborough to witness climate change in the past and present in order to try and imagine the future. Yet, the visual and textual movement between these different temporalities, and across different geographical spaces, works to undermine the seriousness of the issue when considered in relation to the underpinning narrative structure, which follows a linear trajectory identifying the problem, showing the evidence, then offering solutions. Furthermore, the actual solutions proposed are either small scale individual behavioural changes, or large scale technological solutions for the future. A discourse of individualism places responsibility on individuals to make small changes to their lifestyle (Hobson 2006), while political pressure is not even discussed and government responsibility is excised. Scientists are given definitional control, as policy and citizen action (rather than consumer action) are rendered invisible.

Climate action in the 'here and now' – a global problem addressed locally

In contrast to Attenborough's (and thus the BBC's) framing of both the reality and mitigation of climate change as still in question, the focus of Greenpeace campaigns has always been climate action in the here and now. In the early 2000s, Greenpeace focused its attentions upon energy sources, particularly the links between oil companies and politics. Continuing its longer term critique of the activities of oil

Figure 2.8 An individual and collective appeal to climate action from Greenpeace
Source: Greenpeace International 2007.

companies, ExxonMobil became the main focus of campaigning due to its funding of the Global Climate Coalition (which sought to undermine climate science), and its successful lobbying of the Bush Administration in the USA, to remove a USA climate scientist as chair of the IPCC, who advocated governmental action on climate change (Greenpeace International 2002a). From exposing 'dirty' politics, it moved on to promoting renewable energy, particularly wind power, providing both an economic and environmental justification for governmental investment in wind as opposed to nuclear power (Greenpeace UK 2002, 2004). Advocating changes to government policy and industry practices, Greenpeace places individual action within a wider institutional context rather than the individualist discourse offered by the BBC. *How to Save the Climate: Join the Energy [R]evolution* (Greenpeace International 2007) is an explicit call for action through an appeal to a collective movement (see Figure 2.8). Different temporalities are evoked through the word play: revolution connotes abrupt and radical change, while evolution implies a natural, gradual progression. The double meaning simultaneously positions energy changes as a fundamental shift to a new social order, but one that is more developed and natural than the previous one reliant upon fossil fuels.

Significantly, no photographic images are included in the report. Instead cartoon images depict climate processes, climate impacts, climate actions, and an imagined future of decentralised energy production: energy produced and consumed close to its source. Past, present and future are communicated via images, serving to make climate action fun through the humour associated with cartoon imagery; indicating (or assuming) an audience's familiarity with existing climate imagery. All of Manzo's typology of climate images are represented here, however, people figure more prominently than any other type, not simply as victims of climate change but as agents of change. Scientific images in the forms of graphs, computer models of the globe and maps are minimal. The globe itself features here not as a scientific model of a warming world but as an environmental symbol, emblazoned on the chest of the person on the front cover (see Figure 2.8). The inclusion of longitudinal and latitudinal lines on the globe, however, does signify a scientific discourse. As a global call for action, the image draws upon earlier discourses of environmentalism, particularly citizen activism of the 1970s. This global call also evokes an individualist/consumerist approach through branding and logos emblazoned across the chest and the appeal to 'I'. Such contradictory meanings are indicative of the globe as an icon, but also within the context of Greenpeace's use. The slogan (whose signs are intended to be read as 'I love the planet') appears as a motif throughout the report, appearing on a placard of someone protesting outside the Whitehouse (USA), on a T-shirt hanging on a washing line, and as graffiti on an office building. The politicisation of the globe in the form of protest – both legal (demonstration) and illegal (graffiti) – is concurrent with the commercialisation of its environmental meanings through branded clothes. As such, the report addresses a number of different audiences in its attempts to build a collective movement.

Climate science is taken for granted, as attention is focused upon what can be done. Explicitly, humans are addressed as agents of change through an appeal to the human dimensions of climate change: 'This problem is unlike anything in the past. It affects the whole planet and *threatens every living person in every country on every continent*' (Greenpeace International 2007, 2, emphasis in original). This focus upon humans as opposed to animals is reinforced through the text, with repeated references to 'people' and a global 'we'. Through a global discourse, inequity between countries in relation to energy consumption and the ability to cope with climate change is acknowledged as one of (in)justice, where 'every living person on this planet will have to reduce his [sic] personal output of CO_2 to 1.3 tons per annum by the year 2050' (ibid, 6).[2] While some of the changes advocated are similar to the 'small and painless steps' presented in Attenborough's documentaries, others, such as informing oneself, persuading others, and protesting, provide a more socio-political context for action. Although the brochure 'is being published in many countries around the world' (ibid, 3), it is clear that a western neoliberal economic perspective is presented, as the forms of action advocated are possible 'without endangering the global economy and still giving poorer countries the opportunity to develop' (ibid). In the context of energy saving, economic, technological, consumerist and environmental discourses are entwined: 'A lot of household appliances consume much more energy than necessary – even when they are switched off. You can change all of that with a combination of wise purchase decisions and a few simple tricks – and both your *household accounts* and your *climate account* will look a lot healthier' (ibid, 12, my emphasis). The economic justification for climate mitigation, proposed in the Stern Review (2006), enables a language of economics to be used and legitimised in the context of climate change. A neoliberal agenda is articulated through the appeal to 'home-owners' as 'the pioneers' of decentralised energy, although it is not clear how they will achieve this. Cartoons of imagined future suburbs, cities and villages depict the use of solar and wind energy, and the consumption of energy close to its source. Concurrently, the message that 'less is more' contradicts the framing of homeowners as pioneers. The attempt to change perceptions of 'quality of life', assumed in 'the modern world' to be achieved by 'as much, as soon, and as cheaply as possible', does, however, demonstrate an effort to make climate change meaningful through an appeal to, and a questioning of, existing social and cultural values (Greenpeace International 2007).

Discussion – chasing time

Following years of media misrepresentation of the consensus of climate science, the increasing instances of extreme weather events have provided an important news media hook. Although western news media have linked weather events to climate change since the late 1980s (Ungar 1992; Bell 1994; Krosnick et al 2000), the context of scepticism within which these links were made, and thus interpreted, has changed. Where weather events were once criticised by climate scientists and the BBC for misrepresenting climate change, from 2000 onwards, weather has been foregrounded by the BBC and by the climate scientists it has interviewed, as visible evidence of climate change.

---

2 The focus upon human dimensions of climate change is an important shift, signalled earlier in a 2004 report by the International Institute for Environment and Development (IIED) and new economics foundation (nef), that was supported by Greenpeace and other environmental, faith and development organisations/charities entitled *Up in smoke? Threats from, and responses to, the impact of global warming on human development* (IIED and nef 2004).

For the IPCC reports and Attenborough's documentaries, the evidence of climate change is provided by the correlations between observed and simulated data. Computer model simulations and graphs of past, present and (projected) futures are given indexical status through scientific empirical data. These images, and the knowledges produced, are reliant upon clock time, and hence the passing of time, to gain legitimacy. The time encoded in these visualisations is at odds with ecological time, as the observable effects of climate change in the present are the product of past emissions, producing a time lag. Attenborough's documentaries, in particular, are constrained by this sense of time, because it is only when climate change is proven through visible evidence (which includes photographic/video images and computer simulations) that action is advocated.

In contrast, Greenpeace has always maintained that climate change is real, deploying photographs to give it a sense of reality and advocating preventative measures. While Attenborough utilises video images (predominantly) of non-human nature to demonstrate climate impacts, and to provide indexical evidence for the computer simulations, Greenpeace has chosen to use cartoon images in place of photographs to depict the causes, impacts and solutions. Although both Attenborough and Greenpeace advocate 'small and painless steps' for individuals to mitigate climate change, these behavioural changes are framed very differently by each actor. Placed in the context of larger technological fixes, Attenborough renders the mitigation of climate change as the responsibility of individual consumer actions, and large scale technological projects. Greenpeace places the individual within a broader social and political context, advocating changes to behaviour and perception, as well as industry practices and governmental policies. Although both utilise a neoliberal economic discourse, Greenpeace shift attention away from non-human nature to present climate change as a humanitarian crisis.

Whereas the IPCC have been constrained by developments in observational data and computer technology – an issue of economics and time – both Attenborough and Greenpeace place clock time, and the passing of time, as central to climate action. The quantification of time as a means of generating action is intended to communicate, simultaneously, that 'we still have time' (BBC 2006b), but that it is running out. Greenpeace more explicitly identify a timeline for action, stating that 'the next ten years are crucial' (Greenpeace International 2007). Given that the world is already committed to climate change through past emissions, it is interesting to note that adaptation is not discussed. Attenborough's documentary does, however, acknowledge the commitment to increased temperatures when Cox states, 'we don't have very much choice over what happens in the next 25 years because it is largely the function of what we've emitted in the past' (BBC 2006b). Yet, this observation is subsumed within the broader exposition of the documentary and does not facilitate talk about adaptation. In the context of a society based upon clock time, to discuss the need for adaptation to existing climate change does not generate a sense of urgency for action in the present. Instead, attention is focused upon action for the future in the present, as a means of quantifying and containing the threat of climate change. In doing so, ecological time, again, exceeds the boundaries of clock time.

Conclusion

Despite the problems associated with visualising the temporally complex and invisible phenomena of climate change, images have provided an important means of communicating this issue for all three actors, although for very different purposes. In delineating preferred forms of climate in/action images have been used to convey the immediacy of climate change in the present, but also to render it as a future concern. Examined over a period of time, distinct patterns of image use emerge within and across the three actors. For Greenpeace, the indexical function of photographs has consistently been used to substantiate the immediate present of climate change, supported by a textual narrative that asserts climate change in the 'here and now'. Through photographic images of causes (fossil fuels), impacts (weather events, disease) and solutions (solar, wind), the urgency of climate change, and the motivation for climate action, is reinforced through the use of scientific graphs and computer models, whose meanings have been reappropriated and amplified as evidence of climate 'truth'. Originally using a range of images covering the categories of 'Planet', 'Environment' and 'Living Beings' (Manzo 2010), Greenpeace's visual language became more focused upon 'Environment' through distant lands (Arctic/Antarctica) and 'Living Beings' in the form of animals (polar bears) during the late 1990s, in an attempt to create symbolic/iconic environments for climate change. More recently, Greenpeace has moved away from photographic imaging of land and animals towards a renewed focus upon humans as agents of change, including a questioning of existing social values.

For the IPCC, computer model simulations and graphs constitute the basis of climate science and its visual communication. Premised upon the accumulation of observable data in the past to make future predictions, climate models embody a time lag, replicated in the IPCC's consistent appeals for investment in computer technology to offer more credible future scenarios. As such, climate change is communicated as a future issue through the deferral of knowledge production, where investment in future technology remains at temporal odds with the developmental processes of climatic change. As representations which fit into Manzo's category of 'Planet', the computer models map space through cartographic projections that are geopolitical (Jasanoff 2004; Sachs 1999) and distancing. Concerned with influencing policy decisions, in their search for more credible computer model projections, policy action is deferred.

BBC documentaries have consistently appropriated the model projections and graphs of climate science as a means of asserting scientific credibility. In undermining the indexical status of news media images of weather events and reifiying the projections of science as 'truth' (Wolf 1999), the BBC has corresponded with the conservative views of the IPCC, and placed responsibility with scientists and policy makers. While its documentaries have importantly discussed the scientific uncertainties which underpin climate projections, this has been used as a means of appealing for better computer technology to create more accurate climate predictions, thus deferring mitigatory action. Although a shift in representation occurs in the BBC documentary from 2000, as the indexical status of video images of weather events are given credence as evidence of climate change (previously and vehemently denied), it is only through interviews with climate scientists that these images are given credibility. Furthermore, as the BBC documentaries have corresponded to the cautious discourses of the IPCC across this time period, David Attenborough's documentaries from 2006 are conversely incommensurate with climate science, as human induced climate change is posed as a question still requiring answering.

Over the 17-year period analysed, Greenpeace has consistently presented climate change as a future that is present, while the IPCC and BBC documentaries have more explicitly assigned climate change to the future. Although images of science (graphs and computer models) have been deployed in Greenpeace's campaign literature and in the BBC documentaries, they have been appropriated for different purposes, to either amplify or delay the need for mitigating action. Images of climate impacts used by Greenpeace (and by the news media) have been critiqued by BBC documentaries, in favour of the indexical status of scientific images (Wolf 1999). Similar forms of climate imagery have thus been used to convey very different fields of action through their varying contexts of production and reception (Rose 2007). As an institution, the BBC has positioned itself beyond the representational practices of mainstream media, to align itself with the institutional authority of science. Yet, this close alignment with science as a means of conveying its claims to impartiality and honesty (BBC 2009b), conversely denies the political nature of the IPCC, while deferring mitigating action to a future.

Current climate images have been identified as disempowering and disconnected through their simultaneous focus upon distant lands and local impacts (O'Neill and Nicholson-Cole 2009). Over the time period analysed, Greenpeace images have corresponded to this figuration, particularly during the period 1997 to mid 2000s. A more recent focus upon human action through non photographic imaging (Greenpeace International 2007), however, importantly shifts attention onto human agency. The BBC has made climate distanced through its adherence to the scientific images of computer model simulations and the authority of science, but over the last ten years has drawn upon dramatic impacts to humans (BBC 2000) and animal habitats (BBC 2006a, 2006b). Given the BBC's persistent elision of citizen action through its focus upon scientific knowledge however, its more recent use of distant/local images to convey the impacts of climate change must be viewed in this historical and institutional context, which has constrained the promotion of citizen action on climate change.

Significantly, across all three actors, none have discussed the invisibility of climate change, and the temporal disjunctures between its causes and effects, as a means of substantiating action in the present. In focusing upon the visible present (Greenpeace), or investing in future imaging processes (IPCC and BBC), the temporal complexity of climate change and its inherent invisibility has not been mutually recognised.

Chapter 3
Coda: Nature, Vision and Time

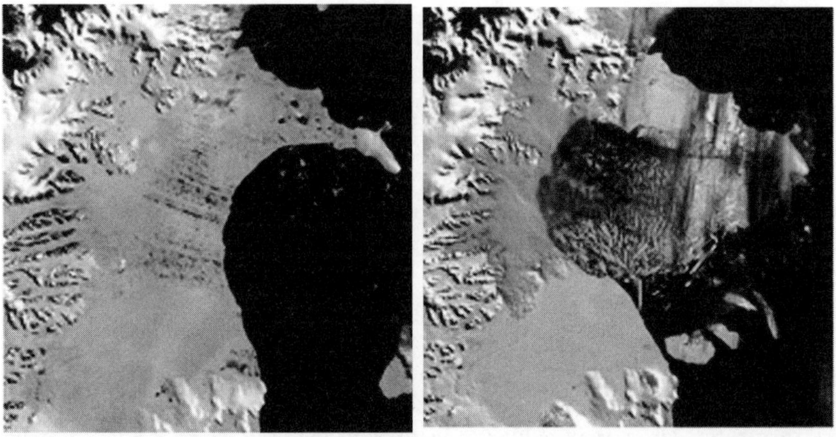

Figure 3.1 Larsen B ice shelf, Antarctica. First satellite image of the ice shelf taken by NASA on 31 January 2002; second image taken on 5 March 2002

Source: NASA 2009.

Captured by satellite imagery, the disintegration of the Larsen B ice shelf in Antarctica in 2002 (see Figure 3.1) recalls the photograph taken five years earlier by Greenpeace who had been the first to observe and document a large split in the same ice shelf (Doyle 2007; see Chapter 1). What Greenpeace first identified and photographed at ground level has been extended by the visual field of satellite imaging, enabling the subsequent collapse of the entire ice shelf to be recorded and made visible from the vantage point of space. Taken from an aerial perspective using a Moderate Resolution Imaging Spectroradiometer (MORIS) sensor (Naranjo 2002), the 'eye' of the satellite selects and focuses in upon this area of Antarctica to reveal the collapse of the ice shelf. Yet, as the ice shelf disintegrates over time, it is only through a comparative analysis of these images, taken at different times, that changes to the ice shelf can be seen. Signifying as 'before and after' shots, the two satellite images are only meaningful in relation to each other. By visually recording particular moments in time in order to show, retrospectively, the passing of time, they map ecological time through industrial time, premised upon visible matter and measured motion in space (Adam 1988a, 1988b). Symbolising a fragile and visible nature at risk, the images also aestheticise nature, utilising discourses of the environment that gain legitimacy through scientific discourse. As such, NASA's satellite images of the Larsen B ice shelf reveal how the interrelated concepts of nature, vision and time in the construction of scientific and environmental knowledge – concepts that have proven to be particularly problematic for the identification and communication of climate change – provide the representational conditions from which the images gain their meaning.

NASA's images gain their legitimacy not only from the indexical nature of the satellite technology used, but also from the knowledge-power relations invested in NASA as a governmental and scientific institution with the economic and technological means to 'collect and record data on a global scale' (Jasanoff 2004, 38). It is both poignant and ironic that the first images of earth from space produced by NASA astronauts are both technologically and imaginatively recalled here in NASA's documentation of the visible impacts of climate change to planet earth. Circulated globally via the internet, I can watch the daily changes to the earth from space via NASA's Earth Observatory listserve (eo-announce@lists.nasa.gov), or choose to ignore it amidst the daily deluge of other emails. Thus, while global satellite technologies can bring an image from space to the personal location of my computer and email account, I am yet further removed from the social realities of climate change. Spectacularly beautiful in their documentation of climatic destruction, such images present a profound disconnect from the environment. Reinforcing the binary distinctions between nature/culture through a visual aesthetics and visual epistemology which privileges sight as a form of knowledge and understanding, they impede our ability to link knowledge of climate change to action.

These two images provide a useful conclusion to the theoretical concerns discussed in Chapters 1 and 2, and offer a bridge to the second section of the book. In the first section of the book, Chapter 1 explored the ways in which scientific and environmental knowledge have constructed, and been constructed through, concepts of nature, vision and time, which have made it difficult for climate change to be made intelligible and culturally meaningful. Through an examination of climate imagery produced by scientists, environmentalists and the media from 1990-2007, Chapter 2 considered how the temporal complexities of climate change have been negotiated through images, and how images have been deployed to create or delay the urgency for mitigating action on climate change. From identifying some of the problems of communication and understanding climate change poses for western knowledge systems and representational practices, the second half of the book now moves on to explore contemporary communication on climate change to consider the ways in which climate change is being made culturally meaningful and relevant, and the extent to which western concepts of nature, vision and time are being questioned or reconceived through contemporary mediations.

PART II
Mediating/Addressing Climate Change

Chapter 4

An Emerging Climate Movement: Questioning Values of Environment, Justice and Faith

Act Fair, Act Fast – Stop Climate Chaos Coalition campaign slogan, 2009
Pray for climate justice – CAFOD campaign slogan, 2009
System change – not climate change – Klimaforum09 declaration, 2009

The Copenhagen conference (COP15) in December 2009 provided an important focal point for many civil society groups and networks campaigning on climate change. As the world's leaders gathered to negotiate a new treaty on global emissions reductions, NGOs and civil society groups from around the world coalesced in Copenhagen (Denmark), to maintain political pressure, disseminate information and commentary, to undertake direct actions and to provide a symbolic presence during the proceedings.[1] Amongst those present were transnational coalitions such as Climate Action Network (CAN) and Climate Justice Now![2] Spanning international, national and local organisations from developed and developing countries, and comprised of environmental, development, faith, justice and youth groups, the conference provided a valuable insight into the diversity of NGOs and networks now involved in an emerging climate movement. While historically, environmental NGOs have been at the forefront of climate change campaigning (Mormont and Dasnoy 1995; Newell 2000), a broader coalition of groups is now addressing this issue. David Meyer has argued that 'effective social movements are defined by broad political coalitions uniting groups with disparate interests and constituencies' (Meyer 2007, 459). The increasing presence of NGO voices from non-environmental organisations thus signals an important shift away from the traditional concerns of environment and nature (as discussed in Chapters 1 and 2) to mobilise a more diverse set of social perspectives and actions on climate change.

---

1 Alongside the official conference, NGOs also established an alternative climate summit, Klimaforum09, which was set up as 'the global civil society counterpart of the official UN conference' (Klimaforum09 2009).

2 CAN is a 'is a worldwide network of over 450 Non-Governmental Organizations (NGOs) working to promote government and individual action to limit human-induced climate change to ecologically sustainable levels' (CAN 2009). Climate Justice Now! is a 'network of organisations and movements from across the globe committed to the fight for social, ecological and gender justice' (Climate Justice Now! 2009).

As representatives of civil society, NGOs seek to influence governmental policies and institutional practices, and to transform social values in order to inspire behavioural change (Wapner 1996; Elliot 1998). Although established actors in the communication and politics of climate change (Newel 2000; Doyle 2007), environmental NGOs have more recently been criticised for being too focused upon achieving national policy regulation at the expense of promoting new social visions and values. In their controversial essay, 'The Death of Environmentalism: Global Warming Politics in a Post-Environmental World' (2004), Michael Shellenberger and Ted Nordhaus argue that environmentalists' conception of the environment as a 'thing' – a separate entity from humans – shapes the forms of climate action imagined and advocated. How we frame climate change – 'how we structure the problem' (Shellenberger and Nordhaus 2004, 14) – affects the solutions proposed, which have problematically focused upon technical solutions and policy regulations (through a pollution paradigm), at the expense of providing more engaging visions for social and political change. Crucially, a radical shift in environmentalist approaches to the 'environment' is necessary in order to build alliances with other groups/industries and to 'craft proposals framed around vision and values' (ibid, 34). Some members of the mainstream environmental movement have responded positively to Shellenberger and Nordhaus through a critical reflection upon their own communicative practices (Green Alliance 2008; WWF-UK 2008a; Crompton and Kasser 2009). Others, however, have offered vocal and sustained critiques of the authors work for focusing upon mainstream USA environmentalism and for failing to acknowledge the history of the environmental justice movement with its commitment to human rights and social justice, thus reproducing a racial, class and gender bias in their assessment of environmentalism (Gelobter et al. 2005; Blain 2005; Brown 2005).

Debates within the environmental movement demonstrate both the significance and the contested nature of social values in the conception and framing of climate change, and therefore in the kinds of actions promoted. While conceptions of 'the environment' as a separate entity from humans need to be questioned (Shellenberger and Nordhaus 2004; Latour 2004), it is equally important to acknowledge the power relations underpinning the various framings and actions on climate change. Indeed, as a wider range of voices from civil society address the issue of climate change, pointing towards more diverse and potentially more meaningful engagements – whether through an appeal to Christian values or to human rights – we also need to be attentive to the forms of power enacted through the discourses and practices of civil society groups and networks working on climate change. Thus, if social values shape action, how is climate change being framed by such a diverse range of NGOs and networks? How do different organisational structures, from international to grassroots, affect the types of values promoted and the kinds of social changes or political actions advocated? This chapter explores these questions through a critical examination of recent climate campaigns from a selection of NGOs and networks, including CAFOD (Catholic Agency for Overseas Development), Friends of the Earth International and Camp for Climate Action.

CAFOD is a Catholic faith based national UK organisation, with links to the global Caritas Internalionalis network in 165 countries (Caritas Internationalis 2009). It campaigns on issues of poverty and justice, mainly in the developing world. Friends of the Earth International (FoEI) is a mainstream environmental NGO, with '77 national member groups and some 5,000 local activist groups on every continent' (FoE International 2010a). Campaigning on issues of environmental and social justice, it is also an international network that grants autonomy to local and regional groups. Camp for Climate Action (UK) is a non-hierarchical and decentralised grassroots climate network working to address the systemic causes of climate change through direct action, education, sustainable living and movement building (Camp for Climate Action 2010a). The different values the NGOs and networks draw upon and use in the framing of climate change are examined in relation to their organisational structure, attentive to the power relations enacted through the symbolic and material practices of climate communication and action. Furthermore, as one of the key arguments of this book is that communication and action on climate change has been limited by the interrelated concepts of nature, vision and time in the production of scientific and environmental knowledge, to what extent and how are these conceptions reinforced, negotiated or questioned by NGO climate campaign strategies? Do the recent climate campaign strategies of NGOs break down or reinforce conceptual distinctions between nature/culture and self/other, which problematically figure humans as separate from the environment?

Before analysing the climate campaigns, this chapter firstly establishes the social and political role of NGOs and civil society groups/networks through existing critical work on social movements. Sutton views social movements 'as resistances to dominant social processes' (Sutton 2000, 38). Although it may be too early to make claims about the existence of a social movement on climate (Meyer 2007), I would argue that the quantity and diversity of NGOs and networks now engaging with climate change does indicate the emergence of a climate movement (Klein 2009). Furthermore, social movements theories provide a very useful set of critical and conceptual tools to examine the relationship between the political opportunities, organisational structure, and the cultural identity of NGOs and networks (McAdam et al. 1996), in the shaping of social values and action on climate change. From a discussion of the various political opportunities and organisational structures of NGOs and groups/networks, this chapter moves on, secondly, to explore the notion of social values as an integral aspect of climate change communication. It examines how these shape the identity and campaigning strategies of NGOs and networks, in turn shaping how the public is called upon to act. As all social values are to some extent exclusionary, presenting a particular set of beliefs at the expense of others, the power relations effected by such values are also interrogated. As such, this chapter focuses upon the cultural dimensions of climate change through an examination of civil society campaigning strategies, an area of research that has been thus far underexplored (Doyle 2009b; Manzo 2010).

The politics and organisation of social movements

Social movements are 'non-institutional' and are often 'separate from more established institutionalised political forms such as pressure groups parties, and the administrative and parliamentary systems of state' (Doyle and McEachern 2008, 84). They characterise a commitment to society and politics from a non-institutional perspective, challenging dominant ideas and relations of power (Doyle and McEachern 2008). This marks a shift from a concern with national politics to relations between the global and local (Princen and Finger 1994), and a move away from state politics to a concern with culture and the struggle for meaning as the basis of collective action. Social movements can thus be usefully characterised as 'a network of informal interactions between a plurality of individuals, groups and/or organisations, the boundaries of which are determined by the collective identity shared by those involved' (Diani 1992, 11, in Pickerill 2003, 19).

While there are differing theoretical approaches to the study of social movements, McAdam, McCarthy and Zald identify three inter-related factors which underpin all of these approaches: 'political opportunities', 'mobilising structures' and 'framing processes' (McAdam et al. 1996, 2). The political opportunities for social movements refers to the 'the importance of the broader political system in structuring the opportunities for collective action' (ibid, 2), that is, the opportunities and constraints presented by national political contexts. Mobilising structures refer to the ways that groups organise themselves, such as formalised organisations or informal networks, and 'through which people mobilise and engage in collective action' (ibid, 3). These organisational dynamics are also structured by different forms of resource mobilisation and how political processes are engaged with. Resource mobilisation includes the kinds of financial, human and time resources available (or those favoured by particular groups/networks), which also constitute a crucial aspect of NGOs ability to influence governments and the public (Diani and Donati 1999). Political processes refer to the ways in which these processes are engaged with, for example, through grassroots mobilisation or more formalised organisational pressure. Finally, framing processes are the 'shared meanings and definitions that people bring to their situation' (McAdam et al. 1996, 5), that is, the ways in which groups forge an identity based upon common values 'or shared understandings of the world and of themselves that legitimate and motivate collective action' (ibid, 6). All three factors are important, but they cannot be analysed in isolation, as the relations between them are crucial to understanding how social movements work. Later on in this chapter, I will examine in more detail the cultural dimension of the NGOs and networks, that is, the framing processes used, which McAdam et al. (1996) identify as being under-researched. Firstly, however, I will discuss in more detail how organisational (or mobilising) structures impact upon the efficacy of NGOs and social movements.

Organisational structures and communicative practices

Della Porta and Diani (1999) contend that most political organisations (including NGOs) are shaped by their resource mobilisation strategies and the efficacy of their political actions. The strategies for resource mobilisation can be attempting to mobilise the largest possible support from the public, or a smaller group of committed activists. The political efficacy of organisations can be achieved through the tactics they use to either disrupt political processes or to integrate them into institutional politics. While some organisation models may favour institutionalised political negotiation over disruptive protest, the choice of tactics is not a property of the organisation and may be contingent. Diani and Donati thus identify two particular dilemmas that organisations face: professional versus participatory, and conventional versus disruptive forms of action (Diani and Donati 1999). Usually, grassroots networks, such as Camp for Climate Action, adopt participatory and disruptive forms of organisation and protest, operating as a decentralised network which may adopt confrontational strategies. In this instance, their greatest resource may be their 'passion and determination' (Corbett 2006, 286). In contrast, established environmental NGOs may assume more conventional and professional forms of action, although these modes may be mixed. For example, Greenpeace is a professional organisation, employing professional members of staff, but it also chooses disruptive forms of protest, which routinely involve local volunteers.

Most theorists have argued that there has been an increasing professionalisation of environmental NGOs since the 1980s, particularly in international and British organisations which have employed a growing number of staff trained in media/communications, law and science, rather than being employed for their experience of activism (Jamison 1996; Diani and Donati 1999; Rootes 2005). While such an approach may reflect government and public support for environmentalism, particularly in Western Europe, such an approach also limits the kinds of participatory action enabled (Diani and Donati 1999). This may have particular consequences for the mobilisation of publics (and the resultant pressure on governments) required for mass action on climate change. Although surveys of public attitudes on climate change in recent years have identified high levels of concern world wide, with exceptions in the USA (Pew Research Center 2009), the equivalent lack of public action on climate change is identified as the attitude–behaviour gap. As international NGOs attempt to understand this gap (WWF-UK 2008a), it may well be that the professional approaches taken by such groups render public participation and mobilisation ineffectual precisely because of their organisational structure, as professional activists may well be deemed by the public to already be doing the job for them.

While environmental NGOs in the north have become increasingly professionalised (Jamison 1996; Doyle and McEachern 2008), since the early 1990s there has been a resurgence of grassroots activism particularly in the UK, through groups such as EarthFirst! and Women's Environmental Network (Wall 1999; Diani and Donati 1999; Anderson 2004). Recent networks such as Camp for Climate

Action (which began in the UK and now takes place in countries such as Australia, Aotearoa/New Zealand and Ecuador) further indicate a resurgence of grassroots activism within the context of climate change, whose activities are often supported and communicated by independent media outlets such as Indymedia (2009). Such a resurgence within a northern context, however, must acknowledge the existing tradition of grassroots activism within the USA environmental justice movement (Sandler and Pezzullo 2007), as well as the predominance of such groups within developing countries, who are often locally based, have women as core members, and put 'bottom up pressure on Third World governments' (Hayes 1999, 239). While the global remit of international environmental NGOs has been criticised for marginalising voices, particularly those of the global south (Rootes 2006), grassroots networks such as Climate Justice Now! are actively supporting activists in the global south.[3]

Marking both similarities and differences across organisations and networks are the communicative practices utilised, and their relationship to mainstream mass media. Since the mid 1990s, the Internet has become an increasingly valuable source of communication and activism for environmentalists (Pickerill 2003; Lester and Hutchins 2009). Within the context of a commercialised news environment, NGO websites are a form of 'self-representation' (Lester and Hutchins 2009, 592), enabling a controlled form of public communication about an organisation or group's message. For large NGOs seeking to obtain public donations within a competitive 'media market place', websites can help forge and maintain 'brand awareness' of the organisation through the way it 'deploys associations, meanings and values to distinguish itself from its nearest competitors in the media marketplace' (Cottle and Nolan 2009). NGO websites thus provide a source of 'contact, interest and information' for the public, as well as 'sources for news journalists in search of content' (Lester and Hutchins 2009, 583). However, the extent to which NGOs seek to provide content that fits the norms and values of mainstream commercial news media, or circumvent mainstream media to provide alternative media messages varies (Pickerill 2003; Fenton 2009), dependent upon the type of organisation. Yet for most NGOs and groups, the Internet and other digital media productively facilitate the building of networks and coalitions across geographical borders (Cammaerts 2007). A recent example of such facilitation is the Klimaforum09 declaration. Supported by over 500 organisations worldwide the declaration was created, negotiated and debated via the Internet (Klimaforum09). Furthermore, as a form of activism, the Internet can be used for political lobbying purposes through email, or as a means of mobilising participation offline (Pickerill 2003), through the

3 The term 'global south' is commonly used to refer to less developed countries primarily located in the Southern Hemisphere. The term 'global north' is commonly used to refer to industrialised and wealthy countries primarily located in the Northern Hemisphere (with the exception of Australia, New Zealand and Japan) (Kegley and Blanton 2009). These terms are also used to designate the unequal power relations between these countries. As such, these terms are open to contestation.

use of text, Twitter and Facebook. Yet as Pickerill reminds us, we must acknowledge the tensions underpinning the use of such technology for both democratic and environmental reasons, as access is constrained by class, gender, ethnicity, language, finance, location and technical skills, and the production/use of such technologies are environmentally damaging.[4]

An emerging climate movement?

A social movement on climate change may occur when organisations and networks work together to 'a form of social and political organisation that typically promotes the broadest range of public communication and political mobilisation' (Meyer 2007, 452). In the months leading up to the Copenhagen conference, a number of mass public demonstrations were held in the UK, where various NGOs, groups and networks came together to create a mutual voice for climate action.[5] The 'Put People First: Jobs, Climate, Justice' demonstration on the 28 March 2009 was attended by 35,000 people, and was co-ordinated to coincide with the G20 summit in London, where the world's leaders from the richest countries gathered to discuss the economic crisis. The demonstration's message was to put an end to poverty and inequality through the creation of sustainable green jobs (Put People First 2009). The coalition of groups was comprised of trade unions, faith, youth, development and environmental groups, which included The National Union of Teachers (NUT), Muslim Council of Britain, CAFOD, WWF and The Trade Justice Movement (ibid).

While some individual NGOs involved, such as WWF, adopt conventional approaches to influence policy at a national and international level, others involved in the demonstrations (though not explicitly linked as partners to the Put People First coalition) take a more participatory and disruptive approach to climate action. For example, the grassroots network, Camp for Climate Action, established a camp outside the European Climate Exchange in London, utilising direct action to critique carbon trading schemes (Camp for Climate Action 2009a). Working together, the organisations and networks constitute a social movement on climate change. When viewed individually, however, they offer very different visions for social and political change which may be conflicting, as some seek to work within the system of national and global politics, while others aspire to creating transformative social changes beyond those enabled by institutionalised political systems. NGOs

4 My intention in this chapter is not to primarily examine the merits of online activism (as undertaken by Pickerill 2003). Rather, my intention is to use the websites as a means of examining current information about each organisation/network, although this does not preclude my consideration of how online activism is used differently by the organisations/networks examined. I am also aware, and acknowledge, that NGOs and networks undertake far more work than can be shown on a website.

5 On 5 December 2009, 50,000 people attended 'The Wave', organised by the UK based Stop Climate Chaos Coalition, with members as diverse as the World Development Movement, Friends of the Earth, The Salvation Army, and the Women's Institute.

and networks involved in the climate movement thus illustrate the complex nature of the contemporary climate movement in relation to competing social and political values.

Framing processes – social values and power relations

> Our actions are deeply embedded in the wider environment, and in the habits and culture and social norms of those around us. They are determined by factors including the search for status, emotions, habits and dominant cultural and social norms. If we are to change, we will do so together (Green Alliance 2008, 11).

Green Alliance call attention to the significance of acknowledging how social norms and values guide our actions. Social values are particularly important when considering the diversity of NGOs and networks involved in climate campaigning and the potential effectiveness of their communicative practices, whether through endeavours to alter people's individual and/or collective behaviours, to affect business practices, or to achieve policy changes. If the climate movement is to engage people's social values in order to effect action and change (Hulme 2008, 2009) then it needs to appeal to a diverse range of people who may hold different social beliefs. At the same time, appealing to a diversity of beliefs may present conflicting views on what constitutes an appropriate set of values to hold. As WWF-UK comment, while a 'successful movement…must be unequivocal in articulating what it stands for' such a recognition 'says nothing about *what* those values should be' (WWF-UK 2008a, 7).

The relationship between conflicting values and modes of action is evident in the debate instigated by Shellenberger and Nordhaus's 'The Death of Environmentalism' (2004), explored further in subsequent publications (Shellenberger and Nordhaus 2007, 2009). Shellenberger and Nordhaus call for a radical reconceptualisation of the environment by breaking down distinctions between humans and nature in order to offer a more positive vision for change. They conceive of new social values as effecting specific forms of economic and political change, specifically facilitating financial investment in large scale research and development, for which they have been criticised. While social values shape the forms of climate action envisaged, these are necessarily contested.

Some actors in the mainstream UK environmental movement have responded to the authors through a serious reflection upon how their own social values influence their communicative practices. In particular, WWF-UK, traditionally focused upon effecting change through policy and business practices, has produced a series of reports examining the role of the environmental movement in a 'post-environmental world' (Shellenberger and Nordhaus 2004). In its *Weathercocks and Signposts* report, climate campaigns which appeal to people's existing lifestyles are criticised for promoting 'small and painless steps' (WWF-UK 2008a, 5) to encourage behaviour change, such as changing lightbulbs, driving less, or buying 'green' products. Such marketing appeals fail to question underlying values of consumerism and individualism, preventing 'more fundamental shifts in policy and lifestyle' (ibid, 2) that are required to address climate change. Campaigns should appeal to 'intrinsic goals' such as 'personal growth, emotional intimacy or community involvement' rather than 'extrinsic goals' such as the 'acquisition of material goods, financial success, image and social recognition' (ibid, 7). Supported by a capitalist economic and political system, such extrinsic values have long been criticised by many grassroots organisations/networks within the global (environmental) justice movement, as well as from indigenous and feminist perspectives (Plumwood 2002; Mies and Shiva 1993; Smith 2007). However, it is of particular interest that the mainstream environmental movement is beginning to re-evaluate its campaigning strategies at the level of social values. Given the longer term criticisms of professional (international) environmental NGOs for their focus upon the north, and the power relations invested in their organisational structures (Elliot 1998), such re-evaluation must be welcomed. The need to articulate a clear climate message necessitates that certain values are agreed across a broad range of organisations and networks.

In order to both encourage and engage people's intrinsic values, WWF-UK has advocated 'identity campaigning' (Crompton and Kasser 2009). Identity campaigning acknowledges the social context through which people's identity – or sense of self – is shaped, which in turn structures our attitudes and behaviours. Crompton and Kasser argue that, 'part of the reason that individuals take on identities that are environmentally problematic is that such identities are supported, and sometimes even encouraged, by the broader social context which, in turn, is importantly shaped by public policies' (ibid, 24). In order to change people's behaviours, campaigns should focus 'on those aspects of a person's identity that either lead them to demand more ambitious change on the part of organisations [government and business], or that underlie their motivation to engage in pro-environmental behaviour' (ibid, 4). Crompton and Kasser argue that environmentally destructive behaviours are often equated with self-enhancing materialistic goals, and that in order to change such behaviours, different values must be encouraged in order to make the required systemic changes: 'If environmental organisations can begin to discuss such values in a more open way and begin to develop policies that enhance them, intrinsic and self-transcendent values will become more legitimised in public discourse' (ibid, 34).

Through identity campaigning, the authors are keen to 'forge alliances across third sector organisations' to promote a consistent set of values (ibid, 65). On one level, their criticisms of materialism fit with the values already advocated by faith groups thus indicating one way in which alliances across groups can be forged. At the same time, however, the capacity for sharing values across groups points towards other power relations which may underpin certain values. For example, Gelobter et al. argue that the 'ideas that drive' building an effective environmental movement and beyond, 'must emerge from a deep encounter between our values, our experience, and the giant social challenges we face' (Gelobter et al. 2005). In order for this to occur we must learn from the civil rights movement of the importance of standing 'with groups that fight sexual and reproductive rights because our histories are so intertwined' (ibid). Thus, if a climate campaign supported by environmental and faith groups promotes the values of non-materialism, but issues of gender, race, class and sexuality are ignored, then the values promoted are undermined.

This chapter now moves on to examine the climate campaigns of CAFOD, Friends of the Earth International (FoEI), and Camp for Climate Action (CCA). Spanning the international, national and local, all had a presence at the Copenhagen conference in December 2009. Through an analysis of their differing organisational structures (mobilising structures) and social values (framing processes) (McAdam et al. 1996), my intention is to explore the ways in which climate change is being made meaningful and the challenges this poses for the political and social efficacy of a developing climate movement.

CAFOD

CAFOD is a Catholic faith organisation working to alleviate poverty in over 60 countries worldwide mainly in the 'global South' (CAFOD 2009a). As a national organisation in the UK working in England and Wales, it is a member of the 'global Caritas network, a federation of 170 international Catholic aid agencies' (ibid). CAFOD works with local groups in other countries, as well as with other UK networks such as the Disaster and Emergencies Committee (DEC), and is primarily funded by the Catholic community in England and Wales. Like FoEI, it is concerned with promoting social justice and sustainable development, although its main focus is upon alleviating poverty. Although CAFOD works with 'people of all faiths and none' (ibid), its ethics of care to humans and environment are explicitly faith based. Conceptions of justice are framed through Catholic scripture: its vision 'is a world transformed to reflect the Kingdom of God', where 'the rights and dignity of every person are respected', 'women and men share equally in shaping their societies' and 'the gifts of creation are nurtured and shared by all for the common good' (CAFOD 2009b). Although dedicated to gender equality, the values which support its organisational structure contradict this commitment. Catholic social teaching (CST), 'based on the core beliefs of Christian faith, such as the dignity of the human person made in the image of God', is viewed as 'authoritative…when set out either at the global level by the Pope or a Synod of Bishops, or at a national level by Conference of Bishops' (ibid). In the Catholic faith, women are not allowed to be a pope, bishop or priest. Organisationally, this top down approach is part of its mobilising structure, reinforcing gendered power relations even as the values of equality are promoted.

A member of both the international CAN and the UK Stop Climate Chaos Coalition, CAFOD is new to climate campaigning, having produced its first climate campaign in March 2009. This followed the statement made by the world's Catholic bishops to world leaders at the Poznan summit in December 2008 of the need to address climate change (CIDSE 2008; CAFOD 2009c). One of the framing processes used in CAFOD's climate campaign is an appeal to justice for the world's poor, as 'those who have done the least to cause the problem' (CAFOD 2010a) are suffering the most. As a form of politicised humanitarianism, alleviating human suffering as a result of flooding, droughts and extreme weather is central to its appeal to western governments to commit to a 40 percent reduction of greenhouse gas emissions on 1990 levels by 2020, and provide funding for the developing countries to adapt to climate change. As Manzo has argued, 'NGOs…typically express their commitments to the vulnerable in terms of *justice* and *solidarity*' (Manzo 2010, 102). For CAFOD, those who reside in the UK (representative of a developed country) are called upon to live 'in solidarity with people who are poor' (CAFOD 2010a) in the developing countries. The discourse of justice for the poor has a moral basis, aligned to climate change through a creationist discourse. The world is 'a gift from God' (ibid), and 'Nature reveals God to us' (ibid). As inhabitants of God's world, 'we' have a duty of care to the earth and all its inhabitants. However, in seeking solidarity, CAFOD, and the Catholic scriptures upon which its work is based, serves to reinforce distinctions between humans and nature, and between humans in the developed and developing countries. Although climate change is envisioned as a question of justice and solidarity, in seeking to address issues of equity and power, CAFOD simultaneously reinforces these through the creation of an 'other' (both human and non-human) in need of protection. The dual discourse of unity and difference is characteristic of what Manzo has called 'geopolitical visions' (Manzo 2010, 96) of the world which construct views of the world from the perspectives of the west. I will explore these claims in more detail later through CAFOD's campaign images and messages.

As an aspect of its mobilising structure, the climate justice campaign focuses upon three forms of action, which are supported through prayer: helping poor communities vulnerable to climate change; motivating the Catholic community in the UK to take action; and calling upon the world's governments, particularly the UK, to commit to reducing greenhouse gas emissions. Actions are conventional and non disruptive (Diani and Donati 1999), comprising of formalised political processes and local campaigning. Campaign materials are accessible via its website, encompassing political lobbying through email as well as suggestions for climate awareness and fund raising events. Explicitly, faith values are used to instigate action: 'Do relate action to faith' (CAFOD 2010b, 1).

Although CAFOD is formally bound by the Catholic teachings articulated by the Pope and bishops, therefore operating a top down approach, many of the climate campaign strategies are realised and enacted at a local level through local parishes and church communities. Actions are centred around the use of existing faith values to engage people, individually and collectively, to address climate change through an appeal to nature. Thus, 'a nature walk or picnic can bring people together to appreciate God's creation and to inspire them to take action', while 'planting seeds or growing vegetables together enables people to reflect on the wonder of creation and on our reliance on the natural world' (ibid). Identity campaigning (Crompton and Kasser 2009) which appeals to the intrinsic values of Catholic belief and community is used to engage people, and while climate campaigning may be a new issue for CAFOD, its use of, and embeddedness within, existing social networks can make it a powerful force in engaging people to undertake action across the span of generations, including children, youths, and adults. The commitment to local action is an import means of engaging people on climate change. Thus, while functioning as a formalised organisation bound by Catholic teaching, it encourages grassroots activities and mobilisation at a local level through the activation of faith.

The Catholic values of an ethics of care to the earth (created by God) and to the vulnerable poor in developing countries is explicitly communicated through the visuals and text of its climate campaign materials and website. Images and symbols are also used to visually communicate and reinforce these values. The image of the right hand figures prominently within campaign materials: as a logo, as a leitmotif within images, and in hand shaped cards to be displayed 'as a symbol of solidarity' (CAFOD 2010b, 1). As cultural symbols, hands can connote friendship, connection, peace, protection and humanity. In the Catholic faith, hands and specific gestures are an integral sign of God's presence and authority: they are used to pray, to make the sign of the cross, to bless and to anoint (Le Gall 1996). The stretching out of hands by the celebrant during a ceremony 'is the ancient posture of the Orant…depicted in the catacombs from the earliest Christian times' (ibid, 72), while the image of Jesus' open hands nailed to the cross is an image of suffering and redemption through faith. The prominence given to the use of human hands in CAFOD's climate justice campaign links climate action explicitly to faith. Yet, the hand also connotes meaning beyond the Catholic faith, signifying as an indexical sign of humans, and therefore symbolically of humanity.

As part of the human body, the hand signifies the human dimensions of climate change in relation to causes, impacts and solutions. Human responsibility for climate action is communicated through image and text, as one of the campaign slogans – 'Our climate is in our hands' – is visually superimposed onto the palm of the hand. In the context of climate change communication, the hand symbol provides an interesting comparison to the sign of a human footprint more commonly used to symbolise climate change. As an indexical sign, the footprint signifies the presence of a human, yet in recent years it has developed as a linguistic and visual metaphor for the negative impact of humans upon the environment, by indicating the production of carbon emissions (see http://footprint.wwf.org.uk/). As something that humans 'do' to the environment, it arguably separates humans from environment. In contrast, the symbol of the hand presents a potentially different relationship between humans and environment. Rather than stepping on the earth, the hand symbolises a human connectedness, implying that through collaboration, we have the power and ability to address climate change.

The human focus of the climate justice campaign is communicated further through photographic images. The focus upon people, or non-human nature, is a significant recent development in the visual communication of climate change (Manzo 2010), suggesting a shift away from animals and landscape as primary climate victims. Yet, as Manzo reminds us, it is important to critically examine how humans are being visualised in climate images, particularly in relation to a discourse of vulnerability, or an 'iconography of disaster' (ibid, 100) which may lead to a public disengagement from the issue. CAFOD's photographic and video images of humans, used on its website and printed campaign materials, depict people in urban and rural spaces in both developed and developing countries. Urban spaces are located in developing countries and depict people campaigning in the form of mass demonstrations (for example, in streets with placards, and at the Bella Centre during the Copenhagen conference). Rural spaces are located in developing countries, where people are imaged walking, standing or working on the land. Thus, while the images do not depict an explicit iconography of disaster through visible impacts to people and land, a discourse of vulnerability through an ethics of care is established through the visual contrast between people from developed and developing countries and the kinds of actions they are doing. The images offer particular geopolitical 'visions of the world' (Manzo 2010, 96), where those depicted campaigning in the streets are responsible for those who are of the land in poor countries; the powerful aiding the powerless, as a voice for the poor is visualised through protest in the UK.

Figure 4.1 Generic CAFOD poster for advertising local climate events
Source: CAFOD 2009.

Through concepts of justice and development, the dual discourse of unity and difference has underpinned western visions of the world (Manzo 2010). CAFOD both reproduces and contests these dualisms in its communication on climate change. In the generic campaign poster used by groups or parishes to advertise their climate event or action, three different photographs of people from around the world appear in circles, linked visually to the other drawn circles in the poster, particularly the large circle in the middle of the hand (see Figure 4.1). Human connectedness is communicated by these interlinked circles, while also representing the globe or 'whole earth' discourse (Cosgrove 1994). The cut out images of people holding hands around the earth further illustrates this global connectedness, reinforced by the visual prominence given to the hand, connoting the human dimension of climate action through the joining of people together in solidarity. Significantly, it is women who dominate these photographic images; a gendered connectedness which represents and reflects the social role of women in the Catholic Church, where those who undertake action are not those who occupy positions of power.

The implied connections also invoke differences and detachment. Firstly, the image of the globe has been shown to signify simultaneously as unity and disengagement, reinforcing a scientific and colonial view of the world as detached and knowable. The gendered connections implied through photographic denotation also connote difference through the kinds of work undertaken, as women in the UK (connoted through the sign of the London red bus in the background) campaign on behalf of the vulnerable women in the developing countries. This illustrates what Manzo has argued is a form of '*innocence-based solidarity* that foregrounds women, children and the elderly in particular (Manzo 2008, 647 in Manzo 2010, 102). Furthermore, such images of gender solidarity serves to both mask and reinforce the real workings of power within the Catholic Church, where women cannot hold positions of authority.

In summary, at the level of social values and framing processes, CAFOD place humans as the victims and saviours of climate change, activated through 'intrinsic values' of faith and moral responsibility. This call to 'intrinsic' values (WWF-UK 2008a, 5) that are anti-materialist and human-centred is positive in the context of engaging people on climate change in a more meaningful way. Similarly the use of existing local networks and communities to motivate people are important mobilising structures that rely upon human and time resources, which have the potential to involve people more meaningfully. The focus upon humans in the context of climate change breaks down distinctions between humans and nature, but conversely reinforces differences between humans based upon the power relations between developed and developing countries. In an attempt to represent solidarity, a sense of othering occurs, reinforced through image and text, as the world's poor are represented as in need of help.

Friends of the Earth International

Friends of the Earth International (FoEI) was formed in 1971, by groups from Sweden, France, the USA and England (Doherty 2006). By 2005, it had grown to 71 member groups in 70 countries (ibid), with over 2 million members and supporters (FoE International 2010a). Calling itself 'the world's largest grassroots environmental network', FoEI 'campaign on today's most urgent environmental and social issues', challenging 'the current model of economic and corporate globalisation' to promote solutions for the creation of 'environmentally sustainable and socially just societies' (FoE International 2010b). Funded by membership fees (12 percent) and subsidies from governments and foundations (87.5 percent) (FoE International 2010c), its uniqueness is that it 'is a federation based on the participation of autonomous national member groups', where there 'is no requirement for members to adopt a particular form of organisation' (Doherty 2006, 864). The national member groups are also linked to regional and local groups. Both Rootes and Doherty contend that of the three biggest international environmental NGOs (including Greenpeace and WWF), only FoEI has succeeded in addressing the issue of a marginalised south through its organisational structure, core values, and work with other groups (Rootes 2006; Doherty 2006). Because of the autonomy of its national groups, 'common perspectives and strategies cannot be taken for granted in the way that they are in Greenpeace and WWF' (Doherty 2006, 866). As such, its organisational structure (as part of its mobilising structure) and core values (or framing processes) are mutually interdependent. FoEI has been campaigning on climate change since the early 1990s, and is a member of a number of climate coalitions including the international networks CAN and Climate Justice Now!, as well as various national networks, such as the UK based Stop Climate Chaos Coalition (Stop Climate Chaos 2010).

Like CAFOD, FoEI frames climate change through a discourse of justice for those 'people of the Global South and Indigenous Peoples worldwide who are most vulnerable to climate change' (FoE International 2010d). Rather than focusing upon monetary compensation from western governments, FoEI shifts the power relation by giving authority to those communities whose knowledge about land and water resources 'is a fundamental element of the sustainable societies we need to build' (ibid). While FoEI acknowledges power imbalances between the global north and south, it seeks to redress these through support of local and traditional knowledges. Rather than campaigning on behalf of communities, FoEI acts as a facilitator to enable community voices to be heard on their own terms, providing 'public and political spaces for affected peoples to share *their* experiences and demand climate justice for *their* communities' (ibid, my emphasis). The recognition and transformation of knowledge/power relations is enacted through its organisational structure and approaches to campaigning, although the international website is constructed for an audience in the north through the use of linguistic terms such as 'their' and 'our'.

Entitled 'Climate justice and energy' (FoE International 2010e), the main climate campaign links climate causes, effects and solutions. Causes are related to 'our unsustainable level of consumption', placing responsibility with people in the global north, while solutions lie in the 'rights of communities to choose their sustainable energy sources and to develop a healthy consumption level', alongside the 'need for greenhouse gas emissions reductions' (ibid). While blame is apportioned, this is not as disempowering to the global south as is implied by CAFOD's appeal for justice for the world's poor. Instead, local and indigenous communities are given credibility for having knowledge of sustainable practices, and for speaking on their own terms. As energy is key to living, framing climate change in relation to energy production and consumption also has the effect of linking all of our actions to climate change, making it more meaningful at an everyday level.

While FoEI advocates autonomous decision-making and empowerment at a local level, like CAFOD it also demands explicit government action, thus engaging with conventional political processes. At COP15, it called for emissions cuts from developing countries and financial aid for developing countries (FoE International 2010f). However, while CAFOD frame this through a discourse of moral responsibility that is faith based, FoEI deploys a discourse of historical responsibility articulated as an economic and emissions debt (ibid). Explicitly critical of carbon offsetting and neoliberalism, it foregrounds a human rights based framework for action, calling upon the United Nations rather than the World Bank to co-ordinate the distribution of financial aid. Although essentially demanding the same actions as CAFOD, FoEI goes further by questioning the institutional structures through which financial aid is distributed, linking to its core values of challenging economic globalisation.

As an international network of autonomous groups, FoEI makes explicit links between the global and local, placing local communities as central to the creation of sustainable societies. On its international website, the climate campaign is communicated through audiovisual testimonies from affected peoples and through written information on grassroots actions from around the world, where each climate campaign is focused upon a local/national context. Human empowerment as the basic for just and sustainable societies is reflected in the imagery of the climate campaigns. Photographs of humans figure prominently, many of which depict groups of people campaigning around the world. Avoiding an 'iconography of disaster' (Manzo 2010, 100), action is communicated through visible activism, as images of climate impacts to humans and non-human nature are minimal. As an environmental network, FoEI conveys, through images and text, its understanding of the environment as human centred, breaking down distinctions between nature and culture, and placing climate change as central to the sustainability of human life and society.

Avoiding the geopolitical visions of CAFOD's representation of the world's vulnerable poor, FoEI manages to reveal unequal power relations while simultaneously empowering those communities most affected by climate change through encouraging grassroots action. It is explicitly committed to 'building a diverse and effective global movement for climate justice' (FoEI 2010g), through the activities of its local groups and through collaboration with other organisations. For example, Friends of the Earth El Salvador helps co-ordinate the 'Movement of Victims Affected by Climate Change (MOVIAC) in Central America' (FoEI 2010h), which positions human survival and the right to life as the central meaning of climate change. Like the core values of FoEI, the movement's critique of global economic and political structures foregrounds the importance of systemic changes in addressing climate change. Because of its organisational structure, which advocates grassroots action at a local and regional level, FoEI cannot take common perspectives for granted (Doherty 2006). Arguably, this makes for a dynamic and more meaningful engagement with climate change as specific political, economic and social contexts inform local/regional/national climate campaigns.

An analysis of FoE UK's Safe Climate campaign illustrates this making meaningful through specific cultural context. The use of the term 'safe' is related to the concept of risk, as climate change is framed as being 'too risky to ignore' (FoE UK, 2010a). This appeal to risk is in the context of recent attacks on the credibility of climate science following the 'Climategate' controversy (Marshall 2009). An appeal to the seriousness of climate change is reflected in the Get Serious About CO2 campaign which calls upon local councils to ensure they cut carbon emissions by 40 percent by 2020, in line with the UK Government's Climate Change Bill (FoE UK 2010b). The focus upon local councils is consistent with FoEI's local and grassroots strategies, as well as its focus upon changing institutionalised politics as people are invited to lobby their local MPs.

Central to the communication strategies of the Get Serious About CO2 campaign is a website that functions as a social networking site, where individuals are invited to upload a photograph of themselves looking serious, and answer some short questions about their level of seriousness and opinion about environmental issues (such as renewable energy, greener homes, planet-friendly farming), and other non-environmental issues (such as tattoos, stinky cheese, reality TV and karaoke) (ibid). The mixing of the serious with the humorous is a strategy that works to engage people in a positive manner. Taking part in the campaign involves supplying your personal details and putting up your photograph with information on your levels of seriousness about the listed issues. As the use of social networking is becoming an increasingly important tool for NGOs and civil society networks (Lester and Hutchins 2009; Fenton 2009), FoE UK's campaign website draws upon the visual discourse of social networking sites where photographic images are central to the construction of identity and community. The website is comprised of a series of photographs which you can click on to find out the name and opinions of that person, with the option of finding out about other individuals with similar interests to the individual profiled by clicking onto a tab which foregrounds a range of selected profiles. Photographs vary from literal portraits, to the creative and abstract, as visual signs of an individual's identity. Visually and structurally the website functions as a form of connectivity, which also utilises a discourse of dating (by clicking on an individual's profile to find out about their interests) as a means of bringing people together.

As an innovative campaign strategy which utilises the discourses and practices of people's everyday lives (in the UK) through social networking, the campaign also draws upon traditional forms of communication, specifically photographs of individuals as representative of identity. This focus upon the individual in relation to an imagined virtual community makes climate action meaningful to a particular audience by utilising familiar visual and social discourses of networked connectivity. However, the purpose of this connectivity is limited to a virtual profile, rather than as a facilitator for community building and mass action in the physical world, indicating the achievement of 'extrinsic goals' such as 'image and social recognition' rather than intrinsic values of 'community involvement' (WWF-UK 2008a, 5). Such an approach is also different to the grassroots activities of FoE El Salvador, demonstrating the differences across FoE groups, as well as the socio-cultural and political context within which groups operate, which informs the kinds of actions advocated.

In summary, like CAFOD, FoEI places emphasis upon the human dimensions of climate change, which links to its core values of creating sustainable and just societies, thereby breaking down distinctions between humans and nature. In contrast to CAFOD, however, humans and communities in the developing countries are given autonomy through their grassroots organisational structure and control over their campaign messages. Furthermore, while justice is a key framing device and core value for the climate campaigns of FoEI and CAFOD, FoEI's is not based upon a limited appeal to faith, but rather a broader humanitarian ethics of justice that seeks equity through addressing the historical responsibility of developed countries to developing countries. FoEI utilises both conventional political and disruptive forms of action, dependent upon the political context of national groups. Similarly, the climate campaigns vary, attuned to particular socio-political contexts.

Camp for Climate Action

The recent development of grassroots climate networks in the UK is epitomised by Camp for Climate Action (CCA). As a non-hierarchical network, CCA links to the history of environmental direct action within the UK (Wall 1999; Diani and Donati 1999; Anderson 2004), although its difference lies in its explicit focus upon the education and practice of sustainable living, realised through its annual, but temporary, week long camp. As such, CCA constitutes a network and an event, whose aims are 'education', 'direct action', 'sustainable living' and 'movement-building' (Camp for Climate Action 2010a). The first camp took place in 2006, near Drax coal-fired power station in Yorkshire, followed in 2007 by a camp at Heathrow Airport, London, and in 2008 at the site of a proposed new coal-fired power station at Kingsnorth, Kent. Each location represents a contested site in the mitigation of climate change, where fossil fuel production and consumption is contrasted with the sustainable living practices embodied by the camp. As a form of direct action, the location of each camp functions as a politicised space (Anderson 2004), providing an important symbolic resonancy (Hansen 1991, Anderson 1997). In 2009, the camp extended its symbolic action to more explicitly represent values of social and class justice, taking place on Blackheath common in London, where 'in 1381, preacher John Ball gave what was probably the country's first speech against class oppression' (Camp for Climate Action 2009b). In 2010, CCA set up camp at the headquarters of the Royal Bank of Scotland in Edinburgh, to protest against its investments in fossil fuels. A series of direct actions undertaken during the annual camp extend its politicised space, and provide a means of further communicating its message to a mainstream audience through its use of spectacular events (Cottle 2008).

The annual week long camp is a form of collective action intended 'to overcome feelings of isolation and helplessness by bringing people together to create a community of resistance' and 'to live in an ecologically sustainable, cooperative way' (Camp for Climate Action 2008a). The camp produces and only uses renewable energy, and all waste is recycled/reused; all of which relies upon the co-operation and commitment of a large group of volunteers. The experience of the camp is then intended as a platform to inspire individuals to adopt these ways of living in their local communities. As part of its mobilising structure, organisationally CCA is non-hierarchical and democratically run. Lacking in major financial resources, it is funded by its members and relies heavily upon human and time resources; all of which help contribute to people's commitment to and ownership of the process. Preparations for, and decisions about, the week-long camp are made through monthly national gatherings run by different local groups across the UK. As a decentralised network of groups with no designated leaders, all participants are considered equal, and decisions are made through a system of consensus: 'In consensus decision making, everyone's voices are heard, and the group tries to find a solution that addresses everyone's concerns' (Camp for Climate Action 2008b).

The grassroots organisational structure is central to the democratic and sustainable values represented by CCA and the social model of living it proposes for addressing climate change. While CCA, CAFOD and FoEI all frame climate change as an issue of social justice, in contrast to CAFOD and FoEI, CCA disregards conventional political processes in the form of institutionalised mainstream politics. Democracy is conceived not through the electoral system and political lobbying, but rather, 'It's up to us, the public acting together, to push solutions that fight against climate change for social justice together' (ibid). As such, the week-long camp, and the processes of consensus decision making through which it is developed, exist as 'a positive example of what the world may be like if we scale back to a more sustainable pace of life' (ibid).

Communication practices are central to CCA's organisational processes and its identity as a grassroots network, utlising a variety of mediated communication to convey its message and to facilitate the camp's organisation. These include a website, short videos, radio broadcasts, social networking strategies such as blogs, facebook, twitter and email/text, more traditional forms of print communication such as leaflets and posters, and face to face communication in the form of regional meetings and the camp itself, which includes workshops and visitor introductions to the camp. Unlike CAFOD and FoEI, the purpose of online activity is not to generate online activism, but to encourage offline activity through involvement in preparations for, and the running of, the camp. The website thus figures as a form of controlled public communication about the network's purpose and activities, used as a means of recruiting and to provide information about regional meetings. Social networking in the form of facebook, twitter, and email/SMS is used to organise meetings, direct actions and co-ordinate the camp. The focus upon social media as a form of networking and co-ordination is a reflection of the grassroots nature of CCA, and its emphasis upon democratic collective action which seeks to bypass the power relations underpinning more institutionalised NGOs.

Broadcasting daily from the camp, CCA generates its own independent media communication through its online Climate Camp TV and radio, corresponding to its organisational structure and identity. It also works to gain access to mainstream media through the production of press releases and through its media liaison team. Mainstream media access to the annual camp, however, is controlled. In 2009, journalists and photographers could only gain access by being accompanied by a climate camper. While this policy was democratically agreed, primarily as a means of ensuring privacy for campers preparing for direct action,[6] it risked being misinterpreted by the media and public as an indication of CCA's mistrust and suspicion of mainstream media. Grassroots networks like CCA must negotiate the tensions between ownership of its own message and seeking to communicate to a mainstream audience through corporate media. This is particularly difficult given the history of its negative press coverage in the UK media. Excessive police tactics at the 2008 Kingsnorth camp, which included the removal of equipment from campers, resulted in many misleading news stories (Longbottom 2008). More so than institutionalised NGOs, grassroots networks like CCA operate in a mediatised culture that is incongruent with the radical social message they seek to communicate, while simultaneously dependent upon the media to convey this to a wider audience.

As a fundamental aspect of CCA's communication strategies, visual and graphic design offer the opportunity for a more controlled form of mass communication (Boehnert 2008). Since the first camp in 2006, the website has become increasingly sophisticated and slick in its visual design, transferred onto the leaflets, posters and stickers it distributes as part of its campaign communication strategies. CCA has worked with Manchester based Ultimate Holding Company (UHC 2008) to produce the campaign designs (ibid). As Jody Boehnert argues, 'Design has helped mediate the public image of Climate Camp and transform the movement into a friendly, understandable, but politically powerful voice of dissent'; an example of 'design activism', where design is used to promote awareness and engage people in a transformative action (ibid). An example of this dual appeal to both mainstream public and grassroots activism is the Swiss Army knife design produced for the 2008 camp at Kingsnorth in Kent (see Figure 4.2). The iconic red sign of the Victorinox Swiss Army Knife (familiar to most western audiences) is connoted by the colour and shape, and through the presence of a logo on the top

6 My thanks to Jody Boehnert from Camp for Climate Action for explaining the reasons for this.

of the knife. However, in this version, the Victorinox logo of a white shield and red cross is replaced with a gold CO2 logo scored through with a diagonal line to signify 'no'. Instead of the usual variety of practical implements Swiss Army knives may possess (such as scissors, knives and screwdriver), in this version the implements include a flower, a corkscrew, a wellington boot, a garden fork, a megaphone, a hand held carrot, a wind turbine, a book, a monkey wrench and a saw. The juxtaposition of these apparently unrelated objects, alongside their altered dimensions (where the wind turbine is the same size as a carrot, as a saw, as a flower), make the image comical, faintly absurd and exceedingly clever.

Figure 4.2 Climate Camp penknife design

Source: UCH 2008.

As an advertisement for CCA, each object is intended to signify as an essential tool for climate action: education, direct action, sustainable living and building a movement (Camp for Climate Action 2010a). Education is signified through the book as a means of learning, but also through the megaphone as a presence for the human voice. Direct action is signified through the presence of other human signs, like the hand and wellington boots, which also connote being on, and working, the land. The products of human labour are food in the form of the carrot, placed in a clenched hand which signifies as a revolution in food production and sustainable living. The monkey wrench draws upon a history of grassroots direct action to imply the potential destruction of property and machinery, and thus illegal action for a common good. Building a climate movement is signified through the sum of the individual parts, where human action will bring people together, and through the literal tools of building and construction – the saw – and the products of construction such as the wind turbine. The objects appeal to a variety of audiences able to read the visual signs of direct protest such as the monkey wrench, but without excluding those who may not be familiar or comfortable with such actions, by including the book and the flower. The cleverness of design is the articulation of ideals through individual signs working together as a whole to communicate action and aspiration in an intelligent, fun and humorous way.

The relationship between its social values and organisational structure is key to the coherency of CCA's identity and the success of its practices. In seeking to build a democratic climate movement based upon the principles of sustainable living, CCA has begun re-examining the diversity of its activists and the values it espouses (Camp for Climate Action 2010b). Reflecting upon the perceived classed division of labour within the camp, with the site 'largely made up of the working class and anarchists while the media and outreach teams was dominated by the middle class/politically liberal' (g.r.o.a.t. 2010, 11), CCA is moving towards a reaffirmation of its radical anti-capitalist and anarchist foundations. At the same time, the need to link with working class movements for solidarity building has become an important call (Jordan 2010). The desire to build a class based movement is linked to the importance of understanding that 'climate change is not an environmental issue…[but] is a system of wider systemic oppression and cannot be addressed without tackling these underlying causes' (anon 2010, 8). In continuously reflecting on the coherency of its own social values and practices, CCA already undertakes the 'identity campaigning' advocated by WWF-UK (Crompton and Kasser 2009) by fostering intrinsic rather than external values, and seeking to build solidarity with other social movements. Its strength lies in its commitment to social change through the creation of a global, grassroots social movement on climate change, which would necessarily bring together a diversity of people and groups who reject 'patriarchy, racism and religious fundamentalism of all creeds' (anon 2010, 6). At present, CCA is primarily based in developed countries (UK, Australia, Aotearoa/New Zealand, Germany and the USA), although its organisational structure does relate to decentralised grassroots movements of the global south. The development of a climate camp in Ecuador in 2009, and in Ghana in 2010, may indicate its expanding global reach beyond developing countries. The difficulties lie in creating a mass climate movement, while maintaining a radical position.

Conclusion

A social movement on climate change would require organisations and networks working together to 'a form of social and political organisation that typically promotes the broadest range of public communication and political mobilisation' (Meyer 2007, 452). CAFOD, FoEI and CCA employ a broad range of communicative and political strategies to engage people on climate change and promote politicised action. While very different in their social values and organisational structure, what binds them together is a shared commitment to social justice and solidarity in the context of climate change: a significant development in climate campaigning which serves to break down distinctions between nature and culture, presenting humans as an integral aspect of the environment. Given the diversity of the three organisations and networks analysed, this is a significant and positive development, indicating strong possibilities for shared campaigning and action on climate change which bring together a diverse range of people.

The differences in social values and organisational structures of CAFOD, FoEI and CCA need to be acknowledged, however, in order to address the power relations that underpin their various framings and action on climate change. For CAFOD, Catholic faith is the framing process through which climate action is encouraged and enacted. Importantly this engages non materialistic and intrinsic values, as advocated by WWF-UK, enacted through strong and existing social networks which engage a wide range of people of different ages. Yet, a focus upon the vulnerable poor in developing countries risks reproducing the geopolitical perspectives of the west (Manzo 2010). The hierarchical organisational structure of the Catholic Church and its teachings also reproduces gendered power relations that are incongruent with a message of justice, assigning the rights for justice to some, but not all.[7] For FoEI, the potential for reproducing similar power relations through its focus upon justice for the people of the global south is countered by its organisational structure, which assigns autonomy to local, regional and national groups, enabling a grassroots approach to develop. While its campaigns are importantly framed according to local, regional or national contexts, FoE UK's Get Serious About CO2 campaign facilitates a surface level engagement with climate change, through its use of social networking discourses which promote an online visual presence rather than offline action. Of the three organisations/networks analysed, CCA's social values and organisational structure are the most aligned, as democratic values of equality and justice are enacted through the organisation of

7 The Catholic Church's condemnation of homosexuality is an example of a position that is antithetical to a concept of justice.

the camp and its vision of sustainable living. Its critical reflections upon how its values are realised through its practices make its identity an ongoing and dynamic process, but one that has to negotiate the tensions between maintaining a radical position and creating a mainstream climate movement.

The development of a climate movement necessitates organisations and networks working together for a common cause. In this context, the role of the mainstream news media in shaping perceptions of climate action must be acknowledged. Reporting on the 'Put People First: Jobs, Climate, Justice' demonstration in March 2009, which included the presence of CAFOD, FoE UK and CCA, a reporter for Channel 4 (UK) television news commented that the power of the protest was diluted because the message was not clear (Channel 4 2009). Furthermore, the broader socio-political context in which the climate movement operates and seeks efficacy, at the local, national or global level, must be considered. While Ed Miliband, the former UK Energy and Climate Change Secretary, has consistently called for a people's movement on climate change (Adam and Jowit 2008; Jowit 2010), the authoritarian actions of the British police in pre-empting legitimate climate protests runs counter to these appeals, making people wary of joining mass public demonstrations (Porter 2009; Jha and Randerson 2009), and rendering Miliband's appeal a token gesture. Concerns over the brutality of police handling of climate protest were also raised at the Copenhagen conference (see Chapter 5 for an analysis of media coverage), while around the world, police intimidation and brutality remains a dominant feature of the policing of protest, routinely exposed by organisations such as Indymedia. In seeking solidarity across a broad range of organisations and networks, a global climate movement can work to counter these forms of oppression and negative media coverage. In doing so, however, it must address the power relations of gender, race, class and sexuality which may underpin its campaign message or organisational structure, else climate justice cannot be achieved (Gelobter et al. 2005).

Chapter 5

Mediating Copenhagen: Communicating Scientific (Un)certainty and the Political (Un)urgency of Climate Action

> What all first-timers to the UN climate process may find hard to grasp at Copenhagen is that this could be the only mass media event in history without a proper beginning or an end, which has no genuine celebrities, no fixed agenda, no guaranteed outcome and is unlikely to throw up clear winners or losers (Vidal 2009, 3).

In framing the UN's COP15 climate conference as a mass mediated event, when compared with the UN's Earth Summit 17 years earlier, environmental journalist, John Vidal, highlights (in the quote above) the mainstream media's increased engagement with environmental issues, and climate change in particular, in both developed and developing countries (Vidal 2009, 3). Vidal notes that while the BBC 'is sending 35 people…China, India and Brazil will be sending nearly 300 journalists' (ibid). Accompanying such a rise in the quantity and breadth of the world media's climate coverage (Boykoff and Roberts 2007) are changes in the forms of mediated communication, with 'every newspaper…sending online journalists, bloggers, video and audio journalists, producers, analysts and Twitterers' (Vidal 2009, 3). Yet, of the 5,000 journalists who received accreditation for attendance, many were representatives of charities, businesses, NGO's, churches and fossil fuel industries, rather than mainstream media (ibid). Furthermore, these figures do not take into account the activities of independent media and activists who produced and disseminated reports, blogs and commentaries from outside the conference centre. Thus, like the proliferation of groups now involved in the climate movement (as discussed in Chapter 4), reporting on climate change has expanded to include a wider range of voices, and across a range of social media platforms.

Despite increased media saliency, or 'newsworthiness', of climate change (Boykoff and Roberts 2007), Vidal's observations, however, also foreground the continuing challenges that the characteristics of news values (Galtung and Ruge 1965; Harcup and O'Neill 2001) pose for climate change reporting in mainstream media. In some regards the Copenhagen conference fits well with existing news values. As a conference of international political significance, attended by '120 Heads of State and Government' (UNFCC 2010), alongside thousands of civil society groups and individuals from across the globe who converged in Copenhagen and set up an alternative People's Summit (Klimaforum09 2009), COP15 corresponds with the events-based nature of news reporting, and environmental news in particular (Anderson 1997; Hansen 2010). The focus upon a powerful elite (Harcup and O'Neill 2001), which included the world's governments and the United Nations, also makes it compatible with the media's preferred news values. Furthermore, the 'Climategate' controversy that emerged in the weeks prior to the conference, meant that media attention was already focused upon the issue of climate science, with the conference thus providing consonance (Galtung and Ruge 1965) or 'follow up' (Harcup and O'Neill 2001, 279) to an existing news story about climate change.

In other aspects, however, the conference did not conform to the characteristics of news coverage, particularly the media's focus upon dramatic events (Boykoff 2009) at the expense of more long term and complex issues. Although characterised as an event, the 12-day conference, arguably, presents difficulties for sustaining coverage and audience interest over this period of time, particularly as the negotiations were complex and involved a large number of political and social actors from across the globe. Apart from prominent politicians such as President Barack Obama (USA), who did not appear until the final day of negotiations, the lack of 'celebrity' figures as a key value in the framing of news coverage (Harcup and O'Neill 2001), and increasingly in environmental news (Brockington 2008; Boykoff and Goodman 2009), makes it potentially difficult to maintain saliency for journalists and readers. Furthermore, as the outcomes of the conference may be ambiguous (as indeed they were) the preferred news value of 'unambiguity' (Galtung and Ruge 1965; Anderson 1997) presents problems for the communication of a complex issue without a clear agenda or results.

Despite the various challenges the conference may have presented for mass media coverage – amidst the competing 'pressures' of an increasing range of social actors involved, and social media forms used, in the reporting of climate issues – mainstream news media continue to be an important actor in the definitional struggles over the production of climate knowledge, and through the courses of action legitimised and promoted via news reporting. Indeed, the Copenhagen conference offers a significant 'critical discourse moment' (Carvalho and Burgess 2005) through which to examine the contemporary mediation of climate change in news media, where two specific events – the 'Climategate' controversy and the COP15 climate conference – temporally and politically converged to focus attention on the scientific and political institutions involved in the politics of climate change. Given the historical importance of print media in the communication of climate change (Bell 1994; Trumbo 1996; Shanahan and Good 2000; Weingart et. al 2000), and its history of 'bias through balance' reporting (Boykoff and Boykoff 2004; Antilla 2005; Andreadis and Smith 2007; Boykoff 2007), an analysis of news coverage of the Copenhagen conference thus provides an important opportunity to critically examine how news media have responded to the political resurgence of 'scientific uncertainty' and consider its potential effects upon public understandings of climate science and science-policy relations.

This chapter offers a critical analysis of UK national newspaper coverage of the COP15 conference, including the week leading up the twelve-day conference. Through an examination of a range of newspapers across the political and social spectrum, including tabloid (*Daily Mirror* and *Sunday Mirror*), mid-market (*Daily Mail* and *Mail on Sunday*) and quality broadsheets (*The Independent* and *The Independent on Sunday*, *The Guardian* and *The Observer*, *The Times* and *The Sunday Times*), the chapter asks: how was the issue of 'scientific uncertainty' in relation to climate science represented, and for what political purposes; how were the various political and social actors involved in the conference framed in news coverage; and how were responsibilities for, and urgencies of, climate (in)action apportioned? While most print media analyses focus upon text at the expense of images, this chapter examines both image and text in the news coverage of COP15.

The focus upon a range of newspapers across the political spectrum is important because of the variety of audiences addressed and the different purposes of tabloid, mid range and broadsheet newspapers. As quality broadsheets 'have an important power of agenda-setting for the public and the other media' and 'are preferred by politicians and other decision makers' (Carvalho 2007, 226), they can influence policy decisions. Although tabloids and mid-market newspapers tend to focus more upon entertainment, celebrity and sports, they do play an important role in making climate change understandable and relevant to people's everyday lives (Boykoff 2008; Boykoff and Mansfield 2008). They also command far higher circulation figures, and therefore potential readerships, than broadsheets. As such, newspapers contribute to making climate change meaningful to the public, as well as informing science-policy decisions through the framing of climate change. In examining newspaper coverage of the Copenhagen conference, this chapter considers the potential effects of news coverage of scientific (un) certainty upon public understanding of climate science and perceptions of the urgency of climate action, as well as the differential roles/responsibilities of (inter) national governments and civil society groups in climate mitigation and action. Before doing so, the chapter firstly offers a brief overview of the purpose of the Copenhagen conference in the politics of climate change in order to provide a context for the analysis of news coverage.

Context of Copenhagen coverage

The COP15 conference was an international political event involving complex and contested negotiations between national governments from 192 developed and developing countries. Two years in preparation, its aim was to forge a legally binding agreement for the reduction of global greenhouse gas emissions to replace the Kyoto Protocol due to expire at the end of 2012. Established in 1997, the Kyoto Protocol set legally binding emissions reductions for developed countries (although the USA did not sign the treaty and were thus exempt), but not for developing countries. The main points of discussion and divergences at the

Copenhagen conference, and leading up to it, were centred upon four areas: the percentage of greenhouse gas emissions reductions for each country and the length of time for this to be achieved (2020 or 2050); the level of global temperature rises on industrial levels to commit to staying below (1.5°C or 2°C); an agreement that developed countries would provide financial aid to developing countries for climate adaptation; and whether to replace the Kyoto Protocol with an entirely new agreement which included developing countries in emissions reductions currently excluded from the Kyoto Protocol, such as China, India and Brazil, or to extend the Kyoto Protocol with the responsibility for emissions reductions remaining with developed countries.

The suggested temperature limits were based upon scientific advice from the IPCC, while differential emissions reductions for developed and developing countries, and monetary aid for developing countries, were premised upon historical inequities, as those countries responsible for producing the least amount of greenhouse gas emissions are being the most immediately affected. As such, the conference illustrates the complexities of climate change as a political, economic, social, scientific, environmental and humanitarian issue. Yet, the conference failed to achieve a legally binding agreement for emissions reductions, financial aid for developing countries, or a decision regarding a successor to the Kyoto Protocol. The outcome of the conference was the Copenhagen Accord, signed by 138 countries, which allowed a voluntary commitment to emissions reductions of an unspecified amount and time span.

COP15 was considered a newsworthy event by all of the newspapers analysed in this chapter, as all provided coverage of Copenhagen, although the extent and depth of coverage varied considerably. Using the original print copies for each newspaper, all news articles, editorials and opinion pieces were analysed, but not readers' letters. The *Daily Mirror* is the second highest selling daily tabloid newspaper (after *The Sun*), with a circulation of 1,225,502 and is the third best selling UK daily newspaper (The Guardian, 2010). Historically, it has supported the Labour Party. The *Daily Mail* is the top selling newspaper in the mid-range, with a circulation of 2,113,134 (ibid), and the second highest selling UK daily newspaper (after *The Sun*). It has traditionally supported the Conservative Party. Of the three quality broadsheets (*The Times, The Guardian* and *The Independent), The Times* is the highest selling, and the second highest UK daily broadsheet (after *The Daily Telegraph*), with a circulation of 521,535 (ibid). It has traditionally supported the Conservative Party. *The Guardian* and *The Independent* are the lowest selling of the broadsheets and of all the national daily newspapers, with circulations of 300,540 (*The Guardian*) and 186,940 (*The Independent*) (ibid). *The Guardian* and *The Independent* provide a liberal, left of centre outlook. While *The Guardian* supported the Liberal Democrats in the UK 2010 national elections, *The Independent* has no declared support for any political party. At the time of the Copenhagen conference the Labour Party were in power (UK).

All three quality newspapers (including Sunday editions) produced extended coverage of the conference making it consistent with a broadsheet prioritisation of science-policy related news. Both *The Guardian* and *The Independent* produced at least two pages of daily coverage prior to, and during, the conference, including a high number of commentary and opinion pieces. The *Daily Mirror* produced more coverage in the lead up to the conference than during it, suggesting that the build up to the event was more newsworthy than the event itself, which failed to sustain interest during the course of the 12 days. The *Daily Mail* had sporadic coverage both before and during the conference, thus deeming the conference the least newsworthy of all the papers analysed. This chapter now moves on to examine how the issue of scientific (un)certainty is communicated by the newspapers.

Framing scientific (un)certainty

Historically, the workings of the IPCC have played a crucial role in international policy decisions on climate change (Shackley and Wynne 1996; Newell 2000; Budescu et al. 2009). Scientific advice from the IPCC thus formed the basis of negotiations at the Copenhagen conference, specifically in relation to achieving international agreement to limit global temperature rises to below 2°C on industrial levels. At the same time, the 'Climategate' controversy which preceded the conference had focused attention upon the uncertainties which are central to the methodologies and future projections of climate science (Shackley and Wynne 1996; Budescu et al. 2009). While these uncertainties do not fundamentally challenge the scientific consensus on human induced climate change, uncertainty can be difficult for 'science-led policy' (Shackley and Wynne 1996, 276), as well as public understanding of science, exacerbated by the traditional assumptions of scientific knowledge and scientific rationality (Adam 1998a; Plumwood 2002; as discussed in Chapter 1) which may hinder discussion about the uncertainties that underpin all scientific inquiry. The email leaks also demonstrated the political context in which climate scientists work and communicate their findings, fearful of discussing the uncertainties of data collection and interpretation in case these are misinterpreted and used to undermine the consensus (Nature 2010). At the same time, the relatively long period of time it has taken mass media to communicate the consensus of climate science, alongside the continuing and disproportionate presence of sceptical voices in the news media (Boykoff 2008), further illustrates the politicised context in which the mediated communication of climate science takes place and the problems encountered in translating scientific knowledge into public understanding (Bucchi 1998).

Mass media are involved in the definitional struggles over scientific knowledge and its representation (Beck 1992). All five newspapers utilised scientific knowledge and expertise to explain, contest and/or reinforce the findings of climate science, and in doing so, actively reconstructed (Carvalho 2007) science according to the different ideological and political positions of each newspaper. The ways in which these different discursive positions are established are discussed in more detail below.

Asserting certainty for policy action

Through its coverage of the leaked University of East Anglia emails, *The Guardian* refutes scientific uncertainty by placing emphasis upon the political activities of the hackers and 'the climate change deniers' who '*seized* on them as alleged evidence that scientists have been manipulating data to support the idea that the world is warming up' (Jha, 2 December 2009, 8, my emphasis). It utilises political actors to discredit the politicised actions of climate deniers, who are viewed as anti-science and, by implication, not rational. Thus, the Labour Government's climate change secretary, Ed Miliband, 'launched a ferocious attack on Conservative politicians who have cast doubt on the science of climate change' and who, 'were acting as "saboteurs"' (Vidal and Carrington, 4 December 2009, 21), while UK Prime Minister, Gordon Brown, led 'a chorus of condemnation against "flat-earth" climate change sceptics who have tried to derail the Copenhagen summit by casting doubt on the *evidence* for global warming' (Carrington and Goldberg, 5 December 2009, 1, my emphasis). As such, science is politicised in order to support explicit policy action at Copenhagen.

Scientific actors in the form of climate scientists and representatives of scientific institutions, including the IPCC, the UK Met Office, NASA, and the Royal Meteorological Society (UK), are similarly accessed to support policy action and to critique it for failing to match the actions required by science. In a front-page headline – 'Copenhagen must fail – top scientist' – NASA's James Hansen, 'the world's pre-eminent climate scientist', criticises the carbon market scheme in favour of substantial cuts in carbon emissions (Goldberg, 3 December 2009, 1). The discourse of scientific rationality which permeates *The Guardian*'s coverage, particularly in the lead up to the conference, is made explicit in an editorial by its sister paper, *The Observer*, in which the gap between science and policy is explicitly foregrounded. While policy failure at Copenhagen is predicted to be related to lack of agreement over financial aid to developing countries and failure to ensure legally binding emissions reduction, it is the 'denial of science' that constitutes a significant problem (6 December 2009, 32). While the 'opponents of a climate deal are newly emboldened by the recent publication of hacked emails', the editorial asserts that 'Man-made climate change is real' and that the 'scientific case for action is *irrefutable*' (ibid, my emphasis). As newspaper editorials act as an 'institutional endorsement' (Carvalho 2007, 231) of a particular view point, the rationality of climate science is asserted as indisputable by *The Guardian* and *The Observer*.

In seeking to advocate political expediency for policy action at Copenhagen, *The Guardian* and *The Observer* thus present climate science as rational and irrefutable, ignoring uncertainties in favour of a rational science based policy. Science is simultaneously depoliticised and repoliticised for the services of climate action – a move which seeks to deny the political agenda of science at the same time as politicising it for policy action. Climate sceptics are given a partial voice – for example, in the article, 'Hellbent on sabotage or just misguided? Meet the climate sceptics' (Randerson, 5 December 2009, 6) – but are presented as irrational, deluded and politically motivated to harm climate policy action.

The Independent similarly seeks to politically discredit the views of climate sceptics, but more explicitly utilises national political differences between the Labour and Conservative Parties to achieve this. In contrast to *The Guardian*'s relative silencing of sceptical voices, *The Independent* gives them voice in order to undermine their credibility as politicians, through their views on climate science. David Davis, former Shadow Home Secretary for the Conservatives, is afforded column space to question the viability of emissions cuts policies, stating that climate change is 'probably' caused by human activity (Davis, 2 December 2009, 9). Significantly, however, this column is preceded, and therefore framed, by a front-page article which announces that leading Conservative figures 'are openly questioning the political consensus on climate change' and in doing so are questioning the authority of Conservative Party leader, David Cameron, who supports policy action to reduce greenhouse gas emissions (Grice, 2 December 2009, 1). In giving voice to a range of Conservative Party climate sceptics, the newspaper seeks to create political differences by discrediting scepticism. In its editorial, *The Independent* presents the sceptical voices as a 'revolt…against the *conventional science* of global warming', to which the paper 'professes itself as rather *baffled* by the climate-change sceptics', rendering their views 'conspiracy theories' (Editorial, 2 December 2009, 30, my emphasis). By deploying a tone of incredulous amusement, the paper renders sceptical views as out of touch and 'eccentric' (Grice, 3 December 2009, 14).

As the conference draws closer, however, *The Independent* deploys a more assertive critique of climate scepticism, calling the inquiry into the leaked UEA emails 'an unwelcome distraction' from the Copenhagen summit, as a means of reasserting the irrationality of sceptics (Editorial, 5 December 2009, 40). An investigation is right, because it not only opens 'the way for those *deluded souls* who still deny there is a man made aspect to global warming to press their case with extra vigour, but because the selectively published emails threaten to undermine the public trust in the *good faith* of climate scientists' (ibid, my emphasis). Furthermore, the 'shrill band of sceptics and their allies, the global oil behemoths…should not be allowed to distract, let alone derail, the historic proceedings at Copenhagen' (ibid). Identifying the politicised tactics of the sceptics enables them to be discredited in relation to the normative views of climate science.

Although *The Independent* (like *The Guardian*) refutes the views of climate sceptics through validating the norms of science, of all the five newspapers it does the most to explain the science of climate change and the history of climatology, including a discussion of the content and political implications of the leaked UEA emails. Rather than dismissing the views of climate sceptics, *The Independent*'s strategy is to reaffirm scientific expertise through the re-presentation of scientific information. Its use of temperature graphs deploys the representational modes of climate science to reassert expertise, supported by the views of climate scientists and representatives of international and national scientific institutions, such as the IPCC, the UNFCC, the UK Met Office and the Tyndall Centre for Climate Change at the University of Manchester. The temperature figures used by the IPCC to form the basis of negotiations at Copenhagen are explained and given some meaning through reference to future scenarios. The importance of achieving a global 2°C temperature limit, 'which is widely regarded as the limit of what the earth can safely withstand', is explained in terms of when emissions should peak, suggested at around 2016 (McCarthy, 10 December 2009, 6), taking into account the impacts of existing emissions.

The temporal disjuncture between the causes and effects of climate change, and between climate science and policy, are acknowledged as a means of legitimising action in the present.

> The global temperature currently stands at about 0.75C above the pre-industrial level, and the carbon dioxide already in the atmosphere is thought to have committed the world to a further rise of about 0.6C – so there is only just over half a degree left before 2C is hit (ibid).

The political commitment to global emissions reductions 'currently on the table at Copenhagen', means that emissions 'will peak far too late for the temperature rises to stay no more than 2C above the pre-industrial level' (McCarthy, 7 December 2009, 4). Although the limitations of climate modelling in providing 'an easy answer to the question of limiting temperature rise to 2C' (McCarthy, 8 December 2009, 4) are recognised, most of the commentary does not explain the uncertainties which underpin the science of climate, such as future scenarios or regional variations.

While reaffirming scientific rationality in order to support policy action at Copenhagen, *The Independent* endeavours to assert further definitional control of climate science by discussing the content and political implications of the leaked UEA emails. Although the attempt to 'hide the decline' in temperatures indicated by tree ring growth after 1960 is 'most damaging', the '"hide the decline" phrase has been seized on by the sceptics to imply that current world temperatures are declining. This is rejected by the scientific community' (McCarthy and Owen, 6 December 2009, 6). The Associated Press investigation into the emails showed 'that the messages don't support claims that the science of global warming was faked', but rather the emails 'show that scientists harboured doubt, however slight and fleeting' (Bornstein, Satter and Ritter, 13 December 2009, 4). Instead of undermining 'the vast body of evidence showing that the world is warming', the emails show that 'the scientists were keenly aware of how their work would be viewed and used, and, just like politicians, went to great pains to shape their message' (ibid). While the political shaping of a climate message is acknowledged, the underlying 'truth' of climate science remains unquestioned.

The *Daily Mirror* similarly emphasises scientific consensus on climate change to support policy action by making very minimal reference to climate scepticism and by amplifying certainty. A sense of urgency in addressing climate change is asserted visually and textually throughout its coverage, as climate change is presented as a fact. Lord Stern, 'chairman of the Grantham Institute on Climate Change and the Environment', states that 'climate sceptics are muddled' (Swain, 2 December 2009, 23). Scientific expertise is reinforced through dismissal of the scientific credibility of climate sceptics: 'Forget the climate sceptics, the real experts are scared: GLOBAL WARMING IS A SCIENTIFIC CERTAINTY' (Swain, 4 December 2009, 40). Here, the representation of 'certainty' is signified through a photograph of floods in the UK town of Boscastle in 2004 and a scientific graph comparing recorded temperature increases and predicted temperature increases from 1985 to 2020. Thus, photographic 'evidence' of climate impacts via weather and temperature graphs showing the similarities between actual and predicted temperature increases, signify both independently and mutually as indexical signs of climate change. Furthermore, the images signify as temporal dimensions of climate change in the past, present and future, reinforcing both their evidential and symbolic force.

The mutual framing of scientific certainty and climate meaningfulness is reinforced through reference to weather – a key strategy of tabloid discourse on climate change (Boykoff 2008). The release of temperature data by the Met Office, who are referred to as 'weather experts', is headlined 'THE HEAT IS ON: the world is at its hottest for 160 years' (Swain, 9 December 2009, 18). The photograph of forest fires in Athens reinforces the evidential nature of climate impacts, and, conversely science, which is represented by a Met Office graph showing temperature rises since the 1850s. Both the graph and photograph thus visually reinforce the symbolic meaning of the written headline. The article states, 'Climate change sceptics claim the world has been cooling since 1998. But the Met Office says the past 10 years have seen a rise of 0.4C above the average between 1961 and 1990…Records go back to 1850s' (ibid). Joss Garman from Greenpeace asserts that, 'The sceptics' case…is not justified by science' (ibid). Scientific 'fact' is established through visual and textual means, where photographic evidence and temperature graphs function to mutually reinforce the 'fact' of climate science and climate impacts. The incontrovertible 'facts' of climate science are also reaffirmed through the institutional voice of the paper, which states that the scientific evidence is clear, so 'Let us hope a deal is reached for the sake of our children and our children's children' (7 December 2009, 8). Scientific certainty is utilised to support policy action, where decisions in the present are presented as affecting future realities. In accordance with the discourse of tabloids (Boykoff 2008), the *Daily Mirror* makes the science meaningful by relating it to everyday lives and concerns. Like the photographs used, this works temporally to make climate change meaningful in the present (our children) and the future (our children's children).

Asserting uncertainty to undermine policy action

In contrast to the assertion of scientific authority in *The Independent*, *The Guardian* and the *Daily Mirror*, a more contradictory view on the status of climate science is offered by *The Times*. It discusses the impacts of the 'Climategate' controversy as a betrayal of public trust and confidence in science. Yet, this betrayal does not engender a questioning of power-knowledge relations between scientists and the public, or the communicative problems encountered in the translation of scientific knowledge into lay understanding (Bucchi 1998). Instead, absolute authority is assigned to the scientists in order to conversely undermine scientific credibility: 'Try as we might, most of us will never understand the science behind climate change. Thus we are forced to trust those who do' (Editorial, 3 December 2009, 2). By establishing such a one-dimensional view of public-science relations, *The Times* places responsibility and blame upon the climate scientists. Yet at the same time, the scientific consensus on climate change is undermined through the presence of a sceptical voice casting doubt on climate consensus. The leaked emails 'strike a blow against the relationship between science and the public, *whether we believe in anthropogenic (man-made) climate change or not*' (ibid, my emphasis). Although 'this newspaper' believes that 'the world's temperature is changing and that humanity is at least *partly* responsible', the '*debate*' is between 'a formidable array of international scientists' and 'noisier, less qualified pundits on the other side', support for whom is growing 'possibly in response to the anti-capitalist and hairshirt mindset who believe that CO2 emissions must be curbed' (ibid, my emphasis). Rather than facilitating a productive questioning of scientific uncertainty to help public understanding of climate science, the undermining of climate consensus is achieved for socio-economic and political purposes, where the neoliberal enterprise of capitalism, and the energy intensive system it is built upon, is presented as the norm. Therefore, while the newspaper can say it believes in human induced climate change, it can simultaneously render efforts 'to curb emissions' as an extreme form of anti-capitalist action.

The contradictory stance of asserting scientific authority in order to undermine it continues throughout *The Times* coverage. For example, a headline, 'The glacier is dying – scientists climb mountain to reach an unwelcome *truth*' (Page, 5 December 2009, 28, my emphasis), is accompanied by the front-page headline, 'Met Office to re-examine 160 years of climate data' (Webster, 5 December 2009, 1). Suspicion is aroused in the Met Office article regarding the validity of climate science and '160 years of temperature data', because of the undermining of 'public confidence in the science of man-made global warming' (ibid). Yet, the issue of public confidence may be seen as a smokescreen for casting doubt on climate data: because the 'analysis of the data will take three years', the 'Met Office will not be able to state with absolute confidence the extent of the warming trend until the end of 2012' (ibid). While the leaked emails may have led to the need for transparency of data and the creation of 'a new and fully open method of analysing temperature data', the article ends by reasserting uncertainty on climate change as the IPCC 'admitted yesterday that it needed to consider the full implications of the e-mails and whether they cast doubt on any of the evidence for man-made global warming' (ibid).

Casting doubt on the science in order to delay or to present a preferred course of action, is reinforced in another editorial the day before the conference begins: 'What none of the Copenhagen participants expected was that even now they would be debating the scientific basis of climate change' (Editorial, 6 December 2009, 20). While the paper declares that 'scepticism about climate change should not make us reject sensible actions', the course of action it promotes is nuclear technology, placing alternative energy as a future possibility 'only when the effect is not to inflict more damage to the planet' (ibid). Although this assertion makes no sense, it works ideologically to link environmentalism to forms of action which are perceived to be economically damaging. The resurgence of government interest in nuclear power as a purportedly 'low carbon' energy source, promotes neoliberal market principles (Doyle 2011) corresponding with those of *The Times*. The paper believes that it is right that 'a dose of scepticism should nip some of the wackier ideas of the climate warriors in the bud' (Editorial, 6 December 2009, 20). Here, the affirmation of scientific doubt is utilised for the purposes of promoting technocratic solutions, such as nuclear power, as a preferred course of future action, whilst simultaneously downplaying the urgency of reducing greenhouse gas emissions. The neoliberal economic agenda through which the climate mitigation actions are framed and judged, is made clearer in the paper's reporting on the proposed expansion of Heathrow airport, which the UK Government's Committee on Climate Change has said 'passes the carbon test', and is in line with Government's emissions targets (Pank, 8 December 2009, 1). Economic expansion is thus the only form of action advocated.

Although *The Times*, unlike the other broadsheets, raises the issue of public confidence and trust in climate science as a result of 'Climategate', it does so in order to undermine climate science rather than facilitate public-science relations. 'Preliminary data' released by the Met Office at the Copenhagen conference, which indicates the last decade as being the warmest on record, 'was an attempt to influence the talks over cutting greenhouse gases' (Webster, 9 December 2009, 16). Yet, this data is viewed positively as a 'back-to-basics measure' which 'cuts out the doubt' because it is based upon 'the simplest most direct measure of temperature – land and sea-based thermometers' (Devlin, 9 December 2009, 17):

> Ice cores, tree rings and supercomputer models. Each year the world of climate science appears to expand in complexity to the point where bystanders must either place their trust somewhat blindly in the conclusions or wonder whether they've been had. The data from the World Meteorological Organisation is a welcome return to basics (ibid).

While the importance of direct and empirical measurements to climate science is made evident, at the same time, the complexity of climate science in relation to predicting future scenarios and regional variations, based upon the accumulation of present and past data, is ignored. In doing so, the temporal dimensions of climate change and climate science are undermined. Instead, climate science and climate change are rationalised through an appeal to the empirical logic of classical science as the most authentic form of science – one which fails to account for the temporal and spatial complexity of climate change (Adams 1998a; Kolbert 2009). Public understanding of climate science is thus simplified and reduced; erasing the uncertainties which form the basis of climate science and of future predictions.

The *Daily Mail* also seeks to cast doubt upon climate science by foregrounding scepticism. However, unlike all the other newspapers, the institutional view of the *Daily Mail* is explicitly sceptical. At the same time, the relative lack of articles discussing either climate science or the Copenhagen conference renders climate change as least newsworthy for the *Daily Mail*. When climate science is discussed, scepticism is amplified in relation to public opinion – '39% of us don't believe that climate change has been proven to be man made' (Derbyshire, 7 December 2009, 10) – and when the economic impacts of mitigating actions will be felt.

More explicitly, scepticism is amplified through the spectre of environmentalism, rather than science. In an editorial entitled, 'Cool heads needed on warming *debate*' (9 December 2009, 14, my emphasis), a debate between environmentalists and 'bemused sceptics' is established rather than between climate scientists and climate sceptics. In doing so, climate science, and specifically climate action, is rendered irrational through reference to the more preferable rational detachment of the 'bemused sceptics'. Similar to the *The Times*, a discourse of scientific rationality is thus deployed to construct climate science as irrational and uneconomic. Therefore, 'if the green lobby is correct', although 'there are many respectable scientists who believe questions still remain about the causes', then 'the cost of supplying 20 per cent of Britain's energy from renewable sources by 2020 will be enormous, holding back growth, health and job creating infrastructure' (ibid). In claiming that 'the global warming issue has descended into ya-boo politics, with the environmentalists seeking to silence all dissent by demonising sceptics' (ibid), the *Daily Mail* asserts the rationality of climate scepticism against the hysteria of climate consensus, displacing the 'debate' onto environmentalists, who are consistently demonised in the discourse of the *Daily Mail* as irrational and anti-capitalist (Doyle 2011; Boykoff and Mansfield 2008). The scientific consensus on climate change is undermined through reference to the '*theory* about man-made global warming' (ibid, my emphasis), while environmentalists promote 'propaganda and zealotry', when what is needed is 'hard science and reasoned debate': an appeal that is certainly not adopted in the discourse of the *Daily Mail*.

Summary – politicising scientific (un)certainty for policy (in)action

All five newspapers deployed the concept of scientific (un)certainty to facilitate political (in)action at Copenhagen, placing science-policy relations centre stage. Within the coverage, science is politicised in a number of ways. *The Guardian*, and to a lesser extent *The Independent*, politicise the actions of the hackers in order to render their scientific claims illegitimate. At the same time, *The Guardian*, *The Independent* and the *Daily Mirror* use political actors in the form of Labour and Conservative politicians to discredit sceptical views (of Conservative politicians) on climate science. These three newspapers also explicitly assert a discourse of scientific rationality and scientific authority through the certainties of human induced climate change to support policy action, similar to research conducted on Swedish newspapers (Olausson 2009). In doing so, they problematically conflate the notion of scientific consensus on climate change with scientific certainty, concealing the uncertainties of climate science to mitigate the potential detrimental impacts of 'Climategate' upon political negotiations at Copenhagen.

Conversely, *The Times* assigns rationality and authority to climate scientists in order to undermine their credibility. By assigning public trust in scientists as absolute, the consequences of 'Climategate' in affecting this 'faith' are foregrounded in order to challenge the legitimacy of climate science, whilst simultaneously amplifying uncertainty in relation to scientific consensus on climate change. In questioning consensus through the concept of uncertainty, policy action is further problematised through the discourse of neoliberalism as carbon reductions are rendered economically unviable. While the *Daily Mail* likewise deploys a discourse of neoliberalism to undermine policy action, and explicitly asserts a sceptical view on climate change, it is through an opposition to environmentalism rather than science that this is expressed.

The Independent discusses the science of climate change in the most detail, including an explanation of the time lags between the production of greenhouse gas emissions and their effects. However, the methodological uncertainties of climate science in relation to future projections, regional variations and the speed of change are not discussed by any of the newspapers. The lack of consideration of the political context in which scientists work (Nature 2010) further limits public understanding of climate science, even as *The Guardian*, *The Independent* and the *Daily Mirror* all assign urgency for policy action through a reaffirmation of scientific certainty. The ways in which the newspapers further assign responsibility for, and urgency to, climate change and policy action is the focus of the next section.

Framing urgency, responsibility and action

Communicating climate urgency – timescales of action

As *The Independent*, *The Guardian* and the *Daily Mirror* all affirm the certainties of scientific consensus on climate change, they similarly assign urgency to addressing climate change through reference to time frames, including clock/calendar time, historical time (such as era/epoch) and past, present and future time. Corresponding to existing news values, all three newspapers create expectation for a coming event by deploying the same running header – 'Countdown to Copenhagen' – which references calendar and clock time. *The Independent* chooses to literally count down the days to the start of the conference, through the subheadings '4 days to go', '3 days to go' etc, while the *Daily Mirror* creates dramatic effect through its subheading: 'FIVE DAYS TO SAVE THE WORLD: COUNTDOWN TO COPENHAGEN' (2 December 2009, 22-23). The mixing of clock time with historical or epochal time is used by all three newspapers to mark the conference as a pivotal moment in human, political and environmental history. For example:

> COPENHAGEN: THE SUMMIT THAT CAN CHANGE THE WORLD – *The Independent on Sunday*, 6 December 2009, 1
>
> Fourteen days to seal history's judgements on this generation – *The Guardian*, 7 December 2009, 1
>
> A historic choice – and only a week to make it – *The Independent*, 12 December 2009, 10-11

UK Prime Minister, Gordon Brown, is quoted in *The Guardian* as saying, 'Copenhagen must be a turning point. Our children won't forgive us if we fail' (Brown, 7 December 2009, 30). Both politicians and journalists thus utilise clock and calendar time to make assertions about historical/epochal time in their attempts to assign urgency to climate action.

On the opening day of the conference, in an editorial shared with 56 other papers worldwide, *The Guardian* utilises time frames to convey the long term and persistent nature of climate change, and the need for political expediency in the present: 'climate change has been caused *over centuries*, has consequences that will endure for *all time* and our prospects of taming it will be determined in the next *14 days*' (7 December 2009, 1, my emphasis). Past, present and future time coalesce to communicate the long term nature of climate impacts and the importance of political action for mitigation in the present.

Reinforcing this temporal urgency for policy action is an appeal to the certainties of climate change in the present and future, through headlines such as:

> Act or face disaster – *Daily Mirror*, 2 December 2009, 22-23
>
> 10FT SEAS MAYHEM TO HIT UK: Climate alert for our coasts – *Daily Mirror*, 10 Dec 2009, 30
>
> RED ALERT: What everyone should know about global warming and the crisis facing our planet – *The Independent*, 2 December 2009, 1
>
> When the sea forces millions from their land – *The Guardian*, 5 December 2009, 7

HOW CLIMATE WILL SHAPE THESE LIVES *The Observer*, 6 December 2009, 18-19

All of these headlines function to affirm the future present of climate change as a means of fostering political action. However, while the *Daily Mirror*'s tone could be said to be alarmist, *The Guardian* and *The Observer* headlines are alarming, and more in line with scientific predictions (Risbey 2007). For the *Daily Mirror*, the certainty of climate change and the necessity for immediate action is also created through images, as photographic 'evidence' of climate impacts to landscape, property and humans reinforces the textual narrative to create dramatic effect. While the sense of urgency communicated heightens the importance of achieving policy agreement at Copenhagen, the relative decline in the *Daily Mirror*'s coverage after the conference begins undermines both the urgency and the reality of climate change and mitigating action, and is thus alarmist. In contrast, *The Guardian* and *The Independent* are consistent in the quantity of their coverage throughout the conference, presenting a sustained sense of importance.

In direct contrast, *The Times* and the *Daily Mail* downplay and undermine the urgency of addressing climate change by presenting policy action as deluded (because there is no chance of success) or unproductive (because it threatens economic growth): a discursive position similar to their framing of science. In a gently mocking tone, *The Times* introduces the conference through the announcement that, 'The world comes together to love the planet' (7 December 2009, 1), then declares in its editorial that 'there is no chance of a treaty that is legally binding' (7 December 2009, 2). The potential for science to help foster policy action is challenged by a continual questioning of its legitimacy, as its response to the UK Met Office release of data showing the last decade to be the warmest on record is framed politically as 'an attempt to influence the talks over cutting greenhouse gases' (Webster, 9 December 2009, 16). Foregrounding the carbon emissions produced by the event, the *Daily Mail* presents the conference as an event that will contribute more environmental harm than good – 'summit hyped to save the planet – more likely to cost the earth' (Derbyshire, 7 December 2009, 10). An apparent environmental concern is thus utilised to undermine long term policy action, as the short term economic impacts of mitigating action are the main concerns of the *Daily Mail*, represented in its headline, 'Climate fight will end cheap flights' (Derbyshire, 8 December 2009, 19). The lack of reflection upon the contributing role of aviation to greenhouse gas emissions indicates the anti-environmentalist discourse of the paper. As such, the temporal urgency communicated by *The Independent*, *The Guardian/Observer* and the *Daily Mirror* are in direct contrast to the lack of urgency and the obstructions to action communicated by the *Daily Mail* and *The Times*.

Assigning responsibility for climate action – political actors

The Copenhagen conference was a political event involving negotiations between the world's national governments. Yet the manner in which political actors and political actions were framed, varied across the newspapers. For *The Guardian* and *The Independent*, a discourse of 'common but differentiated responsibility' (Goldenberg and Vidal, 10 December 2009, 20) is foregrounded. Firstly, a sense of global unity as the basis for political action on climate changes is established through references to 'world community' (The Independent, 8 December 2009, 4), 'international community' (The Independent, 12 December 2009, 10) and 'international unity' (The Guardian, 9 December 2009, 2). These references are supported by headlines such as:

> Will the world finally wake up to the scale of the challenge? – *The Independent*, 7 December 2009, 28

> One world, one agenda – *The Independent*, 6 December 2009, 43

The Independent utilises the discourse of global unity most prominently, particularly in relation to the effects of climate change, by linking the impacts of sea level rises in England (UK) to that of low lying islands and Bangladesh (Connor, 8 December, 6). To some extent, the *Daily Mirror* also communicates a discourse of global connectivity to advocate political action, particularly in its coverage leading up to the conference. Stories about climate impacts in Nepal and the effects of deforestation on the indigenous peoples of the Amazon are followed by stories of impacts to people in Britain through the erosion of coastlines as a result of rising sea levels. These links between the local/national and the global are also established visually through photographs which depict the human subjects of climate change impacts.

As part of the global discourse of unity, *The Guardian* and *The Independent* endorse a discourse of differentiated responsibilities between developed/rich and developing/poor countries. The developed/rich countries are presented as bearing both the historical and moral responsibility for global greenhouse gas emissions. Gordon Brown, British Prime Minister, is figured as the principal negotiator of a financial and emissions reductions deal between developed and developing countries. As the conference progresses without agreement, *The Independent* presents Brown as the diplomatic statesman desperate 'to help salvage deal' (McCarthy, 17 December 2009, 4). However, while the *Daily Mirror* communicates a discourse of global connectedness, it does not acknowledge differentiated responsibilities between developed and developing countries, instead promoting a global discourse of equivalent responsibility.

The Guardian is the only newspaper to articulate and make explicit the unequal power relations between rich and poor countries underpinning the negotiations, and the historical inequity of global resource use. Towards the start of the conference, the assertion of power by developed countries is brought to the fore by *The Guardian* through 'leaked documents that show the world's leaders will next week be asked to sign an agreement that hands more power to rich countries' (Vidal, 9 December 2009, 1). Lumumba Di-Aping, chief negotiator for the G77 group of developing countries, is quoted as saying 'this text robs developing countries of their just and equitable and fair share of the atmospheric space' (ibid, 8). Furthermore, the talks are said to stall 'as Africans accuse developed nations of sidelining them' in negotiations (Vidal et al., 15 December 2009, 17), while the attempt to 'kill' the Kyoto Protocol means that 'Africa is on death row' (ibid).

Despite the foregrounding of power relations, *The Guardian*, along with *The Independent* also reproduce these by criticising developing countries for holding up negotiations for five hours, referred to as the loss of 'precious hours' by *The Independent*. *The Guardian*'s framing of the conference as 'a showdown between rich and poor countries' (Vidal et al., 14 December 2009, 1) fails to account for the inequity of power which underpins this 'showdown'. Furthermore, its environment editor, Vidal, asserts that the need for a global deal takes precedence over the developing nations 'push for 1.5C limit on warming' because 'most industrialised nations have pinned their hopes on 2C' (Vidal, 11 December 2009, 16). Thus, while both newspapers advocate common but differentiated responsibility, they also reaffirm unequal power relations either by not explicating them (*Independent*), or by supporting a global deal of 2°C global temperature limits (*Guardian*) that will effectively consign a large number of developing countries and low lying islands to more extreme climate change.

In contrast to the global, but differentiated responsibility, discourse communicated by *The Independent* and *The Guardian*, both *The Times* and *Daily Mail* seek to undermine the global premise of a climate deal by refuting the responsibility of developed countries to financially compensate developing countries, or commit to larger reductions of greenhouse gas emissions. References to the 'world' occur most frequently to designate 'world leaders', and function as a form of distancing from the event and from responsibility. *The Times* asserts that 'World leaders…are facing the humiliating prospect of having little of substance to sign' (Devlin et. al., 15 December 2009, 4), while the *Daily Mail* claims that 'World Leaders bickered over a deal to save the planet' (Groves and Derbyshire, 19 December 2009, 10). Unequal power relations between developed and developing countries are inverted to present developing countries as failing to equal the commitments of developed countries, and undeserving of financial aid. Disparity between developed and developing countries, which denies historical inequities, is established through headlines such as:

> Developing nations cuts break targets under UK deal – *The Times*, 9 December 2009, 19

> India pledges flexibility but takes a hard line on calls from the West – *The Times*, 4 December 2009, 21

> Carry on polluting, Climate deal tells China and India – *The Daily Mail*, 9 December 2009, 2

Furthermore, *The Times* implies that developing nations have caused the unsuccessful outcome of the conference when it remarks in an editorial that 'tiny Tuvalu caused negotiations to be suspended twice' (19 December 2009, 2).

In relation to financial compensation, the *Daily Mail* most explicitly critiques any form of economic aid by framing it through reference to national politics and the cost to British people; a condemnation that is entwined with that of Prime Minster Brown: 'Brown throws yet more millions at Third World', which will 'drive UK economy deeper into the red' (Groves, 16 December, 2009, 6), making British families suffer. *The Times* also belittles Brown's attempts to forge a financial and emissions reductions deal by arriving earlier than expected in Copenhagen, calling him a 'pyjama man' who 'heads for another sleepover' (Treneman, 15 December 2009, 4). As Conservative Party supporting newspapers which favour neoliberal market approaches, both *The Times* and *Daily Mail* seek to discredit the actions of the Labour Prime Minister, and the aims of the conference, by drawing attention to the economic consequences for Britain of providing financial aid for developing countries. A neoliberal agenda which denies the historical inequity of emissions production is made explicit in an article in *The Times*. Headlined 'Don't let the West carry carbon budget, urge firms', business leaders call upon western politicians not to 'bow to pressure from developing countries' (Davis, 13 December 2009, 9) to commit to larger cuts in emissions as it will harm business.

All five newspapers use political actors to attribute or undermine responsibility for climate action at a national and international level. Deploying a discourse of global unity *The Independent*, *The Guardian* and the *Daily Mirror* call for collective action from the world's governments. *The Independent* and *The Guardian* also assign differential responsibility to developed countries, supporting financial aid to developing countries and greater emissions reductions as a result of historical inequities. Only *The Guardian*, however, makes explicit the unequal power relations underpinning negotiations between rich and poor countries. Both *The Times* and the *Daily Mail* seek to challenge any form of policy action by undermining national politics, and discrediting the claims of developing nations by figuring the developed nations as being unjustly beholden to developing nations. Furthermore, a neoliberal economic discourse presents action as economically unviable.

Assigning responsibility for climate action – social actors

Mass protest and activism constitute an important demonstration of citizen rights in a democratic society, helping create and communicate a sense of collective identity. As Cottle argues, mass protests can help 'overcome possible disconnects between publics, opinion formation and policy-makers' (Cottle 2008, 853), by calling attention to dominant modes of thinking and crystallising alternative political and social realities. The large presence of civil society groups at the Copenhagen conference conveyed the importance of citizen voices in maintaining political pressure on governments, for mobilising publics, as well as offering a range of alternative social visions and values. Cottle reminds us of the importance of mass media representation in (de)legitimising such actions because 'it is in and through the news media especially that the politics of protest and dissent is now generally conveyed to wider audiences' (ibid). In the context of climate change, the mediation of protest in the news media thus plays a crucial role in the construction of civil society groups as (il)legitimate actors in the politics of climate change through the framing of action.

All five newspapers include reports on civil society action at the conference in ways that both legitimate and delegitimate their actions through modes of visual and textual representation. What is apparent is that civil society voices are communicated primarily through visual imagery. Across all newspapers, two forms of images dominate: individual or groups of activists depicted in 'action' (e.g. dressed up, waving placards and banners, shouting); or activists in different forms of 'conflict' with the police (e.g. being held back by the police, or 'attacking' the police). The dominance of visual images corresponds with the events based nature of news coverage, but also the spectacle of protest that is favoured by many civil society groups as a means of gaining media coverage (Cottle 2008). Yet the content of such images, and their textual framing, varies considerably across the newspapers, producing very different views of the role of civil society action in the politics of climate change.

The Guardian and *The Independent* frame civil society groups in a largely positive manner, visually depicting activists in carnivalesque action via costumes and smiling faces, but also as the victims of excessive police force and brutality. This includes images of police with batons raised in an aggressive manner. Such images function as indexical signs which visually demarcate legitimate (activist) and illegitimate (police) forms of action, as well as providing a symbolic presence for civil society. Support is also established through the presence of NGO voices in the textual coverage, via reports on demonstrations and marches, through accessed voices within main articles, and through coverage of daily activist events via *The Guardian*'s Summit Diary and *The Independent*'s Copenhagen Diary. However, while the voices and actions of civil society groups are made visible and framed positively, the range of voices accessed in both newspapers relies upon mainstream international environmental and development NGOs, such as Greenpeace, Oxfam International, Friends of the Earth and the World Development Movement. In doing so, the broad range of networks and groups from developing countries are marginalised. *The Guardian* fares slightly better than *The Independent*, accessing civil society voices more frequently and including the UK Youth Climate Coalition and the Indigenous Environmental Network in its coverage. Although *The Independent* refers to the global network, Climate Justice Action, which seeks to give voice to indigenous peoples and the global south, it labels its members as 'hardline climate-change activists' who are quoted as saying 'we don't want violence, just our voices heard' (Green and Ferguson, 15 December 2009, 13). It discursively links the group with violence through (dis)association. As such, protest from mainstream NGOs from the developed countries are legitimised, while the actions of alternative networks and groups are delegitimised through lack of coverage or via coverage that subtly undermines their approaches.

The *Daily Mail* routinely and consistently excludes civil society groups from its coverage, apart from one reference to the climate march on the 12 December which it frames as 'a weekend of violence' (14 December 2009, 12), supported by a photograph of protesters sitting in lines on the ground 'guarded by Danish riot police' (ibid). Such reportage corresponds with Cottle's characterisation of old style news coverage of protest 'through a dominant law and (dis)order frame, labelling protesters as deviant' (Cottle 2008, 858), which works to discredit legitimate citizen protest. Similarly, the *Daily Mirror* does not access civil society voices in its coverage, also amplifying violence in its reports of climate protest, through headlines such as 'Chaos in Copenhagen' (Yarranton, 13 December 2009, 2) and 'Riot traps police inside summit as climate change talks stall' (Roberts, 17 December 2009, 19). One image depicts clashes between police and protesters, while another shows a black clothed and balaclava wearing crowd of people in the street, with one person holding a firework. A discourse of eco-terrorism is communicated via the balaclavas: a visual sign of terrorism. Throughout its coverage there are no images of activists in carnival style protest, thereby figuring protest as violent and illegitimate.

Although *The Times* acknowledges the presence of civil society groups at the conference, it frames protest in a slightly mocking manner, thus rendering it ineffectual and unnecessary. An image of (benign) theatrical protest and two images of activist interactions with the police – one of which shows a group of protesters sitting in a line outside a domestic house watched over by two bored looking policemen, the other showing a group of smiling police restraining protesters dressed up in leopard costumes – visually depicts the protesters as unthreatening and inept. In contrast, textual references amplify disorder and threat from the 'costumed mobs outside' (Editorial, 19 December 2009, 2) and 'hordes' involved in 'violent scenes' (Leake, 13 December 2009, 2). Spatially removed from the legitimacy of political negotiations *inside* the conference centre, environmental NGOs are criticised for becoming politically involved and 'blurring' the lines between 'diplomacy and environmental activism' by offering support to 'poor nations' dependent upon the likes of Greenpeace and WWF for 'advice and support' (Webster, 21 December 2009, 17). As such, the democratic involvement of civil society groups in climate politics is rendered unlawful and obstructive, and thus public participation in climate action is undermined.

Conclusion

In identifying the Copenhagen conference and the 'Climategate' controversy as a 'critical discourse moment' (Carvalho and Burgess 2005), this chapter sought to explore how notions of scientific (un)certainty were deployed by the print media to support or undermine policy action on climate change at the Copenhagen conference, and consider the effects these discursive positions may have upon public understandings of climate science. The analysis found that scientific (un) certainty was amplified in all newspapers as a means of explicitly informing policy (in)action. Those newspapers that supported policy action (*The Guardian/ Observer, The Independent* and the *Daily Mirror*) downplayed uncertainty by conflating scientific uncertainty with scepticism of human induced climate change and conflating the consensus of climate science with an assertion of the irrefutable certainties of climate science, and thus climate change. In doing so, the uncertainties of climate science were downplayed or ignored, in order to create an imperative for policy action. Furthermore, while the motivations of sceptics were deemed to be political (*The Guardian*), the science of climate change was presented as a depoliticised 'truth'. Those newspapers not supporting policy action either overtly (*Daily Mail*) or implicitly (*The Times*) amplified scientific uncertainty through a discourse of scepticism on climate change and by critiquing the economic impacts of national policy action upon British citizens. While the question of public trust in science was only explicitly taken up by *The Times*, this was done in such a way as to attribute absolute authority to science as a means of casting doubt on its knowledge claims.

The use of science to bolster policy (in)action was facilitated by the sense of urgency communicated and the assigning of responsibilities for action. *The Guardian/Observer*, *The Independent* and the *Daily Mirror* all utilised references to clock, calendar and historical/epochal time in order to communicate the urgency of addressing climate change in the present. Furthermore, *The Independent* explained the time lags, or temporal disjunctures, between the production of greenhouse gas emissions and their effects. Of all the newspapers, the *Daily Mirror* created the most dramatic effect and sense of urgency through the visual amplification of its textual narratives, although its lack of coverage during the conference makes this an example of short-term alarmist discourse (Risbey 2007). In contrast, *The Times* and the *Daily Mail* sought to downplay the urgency of climate change by challenging the science and the political efficacy of (inter)national governmental action at Copenhagen.

While science is utilised to facilitate policy (in)action, politicians and civil society groups and their representatives are accorded different responsibilities. *The Guardian*, *The Independent* and the *Daily Mirror* deploy a discourse of global unity, although *The Guardian* and *The Independent* also advocate differentiated responsibilities for developed and developing countries as a result of historical inequities in greenhouse gas emissions. Yet it is only *The Guardian* that explicitly discusses the unequal power relations between rich and poor countries underpinning the negotiations. The representation of civil society groups by *The Guardian* and *The Independent* further reinforces the necessity of action through visual civil protest and through condemnation of police tactics. Conversely, the events based coverage of the *Daily Mirror* means that it presents protest in a negative manner, heightening violence through image and text and delegitimising civil society action in favour of science and policy action.

The Times and the *Daily Mail* choose to dismiss the global premise of a climate deal, instead gently mocking the attempts by the world's politicians to forge an agreement (*The Times*), or inverting power relations by casting developed countries as unjustly treated by developing countries (*Daily Mail*). The mocking tone deployed by *The Times*, which works as a form of distancing, is extended to civil society groups who are viewed as ineffectual and unthreatening. While the *Daily Mail* routinely excludes civil society groups from its coverage, it frames protest as violent and deviant, conversely blaming environmentalists for the conference's failure as a result of acting too politically. Explicit boundaries between politics, science and the public are thus advocated, yet the *Daily Mail*'s coverage of the conference fails to assign responsibility for any of these actors, thereby legitimating inaction.

Of concern for public understanding is the systematic denial or undermining of climate change and climate science by the *Daily Mail* and, though to a lesser extent, *The Times*, which represent a significant majority of the UK daily newspaper circulation. The silencing of a legitimate voice for civil society in these newspapers simultaneously places responsibility on politicians and scientists. At the same time, the efficacy of these actors is undermined in newspaper coverage of climate change politics. As such, democratic participation is erased by the simultaneous denigration of all three actors as variously inept or illegitimate.

This exploration of newspaper coverage suggests implications for public understanding of climate science and the assigning of responsibilities for action. Given the context of 'Climategate' and the Copenhagen conference, it is understandable that liberal newspapers would seek to relegitimise climate science by reasserting certainty, particularly given the historical difficulties in communicating the scientific consensus on human induced climate change within mass media. However, an opportunity was missed to properly explain the uncertainties which underpin climate science, but which do not themselves negate the consensus of climate science. A discussion of the meanings of the language used by climate scientists and the IPCC in relation to terms such as 'likely' or 'highly likely', and the percentages of certainty these represent, could have generated better public understanding of the notion of certainty and uncertainty as used in climate science (Shackley and Wynne 1996; Budescu et al. 2009).

Chapter 6
Sustainable Consumption?: Reframing Meat and Dairy Consumption in the Politics of Climate Change

> [S]uccess is likely to come from persuading mainstream consumers that in adopting lower-carbon lifestyles they can save money, have control, look good, do the right thing without being an environmentalist and still be themselves (IPPR 2009, 3).

> Preventing the collapse of human civilisation requires nothing less than a wholesale transformation of dominant cultural patterns. This transformation would reject consumerism – the cultural orientation that leads people to find meaning, contentment and acceptance through what they consume – as taboo and establish in its place a new cultural framework based on sustainability (Assadourian 2010, 3).[1]

As profligate users of cheap fossil fuels and natural resources, capitalist consumer societies have been the primary cause of global increases in greenhouse gas emissions over the last 150 years, and the last 50 years in particular (nef 2009a). High-income, middle-class lifestyles contribute the most to global greenhouse gas emissions, with 7 percent of the world's population responsible for 50 percent of the world's emissions, while 'the poorest 50 percent of the global population…are responsible for just 7 percent of global emissions' (nef 2009a, 3; WorldWatch Institute 2010). Such inequities are further revealed by calculations based upon current rates of consumption which posit that if the world's population consumed at the same rate as Europeans, then three planets' worth of resources would be needed, and five planets at the rate of North Americans (WWF International 2008; nef 2009a). Consumption practices include the direct consumption of food, water, energy and construction materials, as well as indirect consumption through 'extraction, production, disposal and transportation of goods' (Wilk 2002, 5). In order to mitigate climate change, consumption levels in developed countries need to be drastically reduced (WWF-UK 2008b; nef 2009a; WorldWatch Institute 2010).

As the opening quotes to this chapter demonstrate, however, there are conflicting views as to how to reduce the carbon intensity of consumption practices; ranging from an outright rejection of consumerism, to a revisioning of consumer

1 Source: Worldwatch Institute, State of the World 2010: Transforming Cultures, www.worldwatch.org.

lifestyles to encourage the adoption of lower carbon practices which do not disrupt an individual's sense of self. Such approaches can be distinguished as an 'ethics of consumption', which is largely critical of consumer society and seeks to reduce levels of consumption, as opposed to 'ethical consumption', where consumption is seen as 'a medium for moral and political action' through 'consumer boycotts', 'fair-trade campaigns' or purchasing products deemed less environmentally damaging (Barnett et al. 2004, 21).

Theorists of contemporary consumer societies have shown that consumption plays a major role in the construction of identity, and has become the primary means through which social relations are formed (Miller 1998; Bauman 2007; Soper 2008; Gilbert 2008; Littler 2009). This chapter explores the role of consumption in the production of consumerist lifestyles as contested sets of practices integral to the politics of climate change. More specifically, it examines the role of food consumption, in the specific form of meat and dairy products, as an important site for the mitigation of climate change. Through an analysis of recent public communication campaigns from civil society groups and individuals, including Paul McCartney's Meat Free Monday (UK), Peta2's Meat's Not Green (USA) and Friends of the Earth UK's The Food Chain Campaign, this chapter examines how meat and dairy's contribution to climate change is made visible through the campaigns, the discursive tactics used to address the consumer/citizen, who is being made responsible for changing consumption practices, and how the social norms of food consumption are being questioned.

While there have been some analyses of climate campaigns focusing upon reducing household energy consumption, particularly Government campaigns (Slocum 2004), the promotion of public campaigns which focus upon the role of industrial meat and dairy production and consumption in the mitigation of climate change has received far less attention. Given the impact that the livestock sector has on global greenhouse gas emissions, contributing 18 percent of global emissions according to the UN (FAO 2006), this is a significant oversight. It illustrates how forms of consumption are embedded within existing socio-cultural practices, where meat and dairy consumption is largely conceived as a 'natural' practice within western, and middle-class, societies (Heinz and Lee 1998). Governments have been disinclined to promote the reduction or elimination of meat and dairy because they are generally 'reluctant to tackle questions of personal choice and consumption', instead focusing their climate campaigning efforts upon household energy consumption (Robins and Roberts 2006, 39). Altering consumer practices, particularly those that are experienced as both 'natural' *and* aspirational – as meat consumption rises in middle-class societies – therefore requires acknowledging the symbolic dimensions of food consumption, as well as the social, economic and political systems which support food 'choices'.

In the context of consumer culture, meat and dairy products are commodities: a set of resources to be used by humans, which 'corresponds to the commodification of nature' (Plumwood 2002, 144). Campaigning efforts that identify the relationship between climate change and industrial meat and dairy production – and the consumption of its commodities – are important as a means of revealing its significant contribution to global greenhouse gas emissions. They also help us to question practices of commodification which view nature as a set of resources to be used and exploited – both of which demand a reconceptualisation of meat and dairy production/consumption as unsustainable and unethical. The ways, and extent to which, the campaigns analysed here contribute to this reconceptualisation of industrial meat and dairy production/consumption, within the context of unsustainable consumer culture and climate change mitigation, is the focus of this chapter.

While an overall reduction in consumption is required to address climate change, promoting an 'ethics of sustainable consumption' (Michaelis 2006, 335) necessitates both revealing and changing the energy-intensive and unethical/unjust processes of many contemporary consumer practices undertaken routinely. What makes the campaigns of interest for an ethics of sustainable consumption is that they represent a rejection and revisioning of certain consumer practices, utilising, to varying degrees, both an 'ethics of consumption' and 'ethical consumption' within their campaign strategies (Barnett et al. 2004, 21). Given the highly symbolic nature of meat (Adams 1990; Lupton 1996; Heinz and Lee 1998; Allen and Baines 2002) and dairy consumption, alongside its key role in mitigating climate change, exploring both the possibilities and limitations of the various approaches to changing meat and dairy consumption practices is of real importance to the goal of reducing consumption levels within developed countries (WWF-UK 2008a, 2008b; nef 2009a; WorldWatch Institute 2010), as well as making our consumption practices more equitable and sustainable. Before analysing the campaigns, this chapter first engages with academic literature on (sustainable) consumption drawn from cultural and media studies, cultural geography and the social sciences, in order to frame the analysis within the context of debates on the role of consumption and consumerism within the politics of climate change.

Consumer societies and the concept of sustainable consumption

> Unlike *consumption*, primarily a trait and occupation of individual human beings, *consumerism* is an attribute of *society* (Bauman 2007, 28).

In seeking to distinguish consumption from consumerism, Bauman foregrounds the ideology of contemporary consumer societies. While consumption is an individual act, which involves the acquisition of goods and services, consumerism is 'a type of social arrangement that results from recycling mundane, permanent and so to speak "regime-neutral" human wants, desires and longings into the principal operating force of society' (Bauman 2007, 28). Consumerism is an ideology which structures both individual and social relations, where human desires, experienced as needs, are expressed through practices of consumption. Although capitalist consumer societies produce more goods than are materially needed, the notion of human need is central to a consumer society, where basic material needs, such as food, warmth and shelter become entwined with the concept of human well-being and identity that gets equated with material consumption (Michaelis 2006). Human needs and human wants/desires become increasingly difficult to distinguish because 'material consumption is a major route to belonging to a community and achieving status within that community' (ibid, 330). As such, consumption cannot be extricated from consumerism, but rather, acts of consumption are reified to become the primary basis through which social relations are lived and experienced. Bauman argues that 'the secret of all successful "socialization" is making individuals *wish* to do what is *needed* to enable the system to reproduce itself' (2007, 68). Consumerism is thus both an economic and social activity through which an individual's sense of self and community is embedded.

The ideology of consumerism promotes the notion of individual consumer rights, where 'individual freedom to own property and to consume it is taken to be a fundamental right of all human beings' (Michaelis 2006, 330; Gilbert 2008). As Andrew Szaz observes, 'individualism as a mode of experiencing and as a mode of action, is at the core of our culture. Consumption is too. Consuming occupies much of our time, attention, enthusiasm – passion, even' (Szaz 2007, 7). The production of goods and services requires the use of energy and resources for their manufacture, distribution, consumption and disposal. Yet, modern consumer society renders invisible the energy (or carbon) intensive and profligate system which sustains it. The social legitimacy of the desire (or 'need') for products and services supercedes the reality of unsustainable amounts of energy and resources used for the production, distribution and consumption of consumer goods (including food products). The space-time distanciation between the manufacturing of goods and their consumption contributes to the fetishisation of commodities (Marx [1867] 1990) by concealing the commodity chains, or the labour, involved in the production and distribution of goods. Furthermore, the obsolescence built into material goods, where products are designed to last only a couple of years, creates the material conditions through which consumer desire is perpetually stimulated and reinforced (Chapman 2005).

Cultural theorists have identified and explored a range of 'alternative', 'ethical' or 'radical' forms of consumption (Bryant and Goodman 2004; Harrison et al. 2005; Littler 2009). These include changing unethical economic practices through fair trade (Bryant and Goodman 2004); promoting green consumption via the production/consumption of environmentally less damaging products such as organics or recycled clothing/fashion (Fox 2009); boycotting corporate brands (Klein 2001) or specific products that are environmentally harmful (Greenpeace International 2010); seeking to change consumer practices through buying local produce (Guthman 2008) or the Slow Food Movement (Slow Food 2010); or via 'anti-consumer activism' such as 'Buy Nothing Day' and Bill Talen's 'The Church of Stop Shopping' (Littler 2009, 70, 80). In the context of climate change, efforts have been made to introduce systems of quantification which identify the amount of greenhouse gas emissions generated by food produce (often referred to as food miles), although this is notoriously difficult to quantify given the complexity of food production, packaging, transport and modes of consumption (Hertwich and Peters 2009).

Many alternative forms of consumption target affluent, middle-class consumers, reinforcing the economic and social logic of consumer culture (Szaz 2007). Rather than dismissing such consumer practices as inherently classed, however, Kate Soper draws upon and utilises class distinctions to advocate a new anti-consumerism called 'alternative hedonism' (Soper 2008, 567). Arguing that consumption by the affluent elite is 'unlikely to be checked in the absence of a seductive alternative', Soper calls for a 'new erotics of consumption' (ibid, 571) based upon the pleasures of not consuming. Both altruistic and self interested, such alternative hedonism would advocate the pleasures of 'less tangible goods such as more free time, less stress, more personal contacts, a slower pace of life' (ibid, 576). Such a vision is certainly enticing given the fact that it is the global affluent on middle to high incomes who contribute the most to greenhouse gas emissions (nef 2009a; WorldWatch Institute 2010), and who must be the target of campaigns to promote less consumption. The focus upon pleasure rather than displeasure is also welcome given the failure of existing climate campaigns, based upon generating fear, to engage citizens productively (O'Neill and Nicholson-Cole 2009). Soper's proposition, however, reinforces class distinctions by placing power with an elite group of consumers who already have the material and symbolic means to create distinctions through a politics of taste (Bourdieu 1984). Locating power in the choices of the affluent consumer perpetuates the individualist premise of consumerism.

Engaging with the notion of sustainable consumption, rather than green or ethical consumption, may prove more productive for thinking about the embeddedness of human actions within the environment. There is much debate, however, over what the term means (Elliott 2006; Robins and Roberts 2006). First defined by WCED as part of the Brundtland Report of 1987, sustainable development is 'economic and social development that meets human needs now without compromising future generations' ability to meet their needs' (Hobson 2006, 305). In this definition, sustainable consumption means taking into account the environmental impacts of consumption, whilst also meeting the needs of individuals in providing a better quality of life. But as Kersty Hobson observes, this approach represents 'the "rationalization" of lifestyle practices' (Hobson 2006, 305), also known as ecological modernisation (Hajer 1995), where science, technology and efficiency are invested with the power to address environmental problems, and the practice of consumption is not threatened. Structural changes are not sought, but instead new preferences are incorporated 'without impinging upon individual's (supposedly) sacred and deeply entrenched lifestyles' (Hobson 2006, 307). Part of this approach is the assumption that as more information is given to individuals about the environmental impacts of certain products, then 'individuals will change their consumption behaviour' (ibid, 306). In figuring the individual as rational, such an approach ignores the social and habitual nature of consumption practices which can be difficult to change (Clarke 2008). Furthermore, in placing emphasis upon 'individual consumption practices' that remain 'positioned within contexts and infrastructures not conducive to living sustainably' (Hobson 2006, 311), the conditions and incentives for policy regulation are left unaddressed (Robins and Roberts 2006).

The sustainable consumption of food, goods, resources and energy must be a goal, which takes into account the current inequities of global consumption through a radical reduction of consumption in wealthy countries (Rees 2008; Seyfang and Paavola 2008). To change consumer practices within wealthy societies, cultural and social norms need to be altered to meaningfully reveal the ecological unsustainability of many of the behavioural and consumer practices undertaken on a daily basis. Researchers have already revealed that increased consumption does not lead to a proportional increase in happiness (nef 2009b; Rees 2008; Jackson 2008). At the same time, we also need to acknowledge how our daily practices are embedded in social networks which may make it difficult to change certain behaviours (Hobson 2003). Whilst in agreement with WWF's assertion that the 'ultimate imperative for society [is] to consume less rather than differently' (WWF-UK 2008a, 36), I contend that consuming differently also has an important role to play in addressing climate change, because it facilitates a questioning of social and cultural norms about consumption, as well as making climate more culturally meaningful by relating it to our sense of self – our identity – through our routinised daily practices.

This chapter now moves on to examine consumer/citizen based climate action through an analysis of civil society campaigns focusing upon meat and dairy consumption: Paul McCartney's Meat Free Monday (UK), Peta2's Meat's Not Green campaign (USA) and FoE UK's The Food Chain Campaign. All of these campaigns endeavour to link meat and dairy consumption to the production of global greenhouse gas emissions and climate change. Each campaign focuses its message upon different aspects of livestock production and consumption, revealing various *processes* implicated in the meat/dairy commodity chain, and addressing audiences in diverse ways. Food consumption is more intimately linked to subjectivity and identity (Lupton 1996) than other external forms of energy consumption through behavioural activities such as driving, cooking, lighting and powering electrical appliances. Given the criticisms already discussed regarding the focus upon the individual consumer at the expense of collective and policy action, this chapter examines how the campaigns locate an individual's food consumption practices in the context of wider structural conditions. At the same time, it examines the extent to which the promotion of behavioural change through meat and dairy consumption invites small or more radical ethical changes through its appeal to a sense of identity and community, and its questioning of processes of commodification in the context of climate change.

Reframing vegetarianism and veganism – the climate politics of meat and dairy consumption

> Go vegan, fight climate change – Animal Aid campaign slogan, 2006

According to the UN the global livestock sector 'is one of the largest sources of greenhouse gases and one of the leading causal factors in the loss of biodiversity'

(FAO 2006, 267). Emissions are predominantly produced in the use of land for livestock pastures and for animal feedcrops, both of which lead to significant deforestation which is a leading contributor to climate change (ibid). The digestive fermentation of plant proteins by livestock produces methane via animal effluence, and animal waste produces nitrous oxide: both methane and nitrous oxide are significantly higher contributors to climate change than carbon dioxide. Furthermore, as irrigation is required for feedcrops grown for animal feed, 'Rearing animals for food uses large areas of agricultural land, vast quantities of water and significant amounts of energy' (FoE UK 2008, 6). Soybeans are the main crop grown for animal feed, and '97 per cent of the soymeal produced worldwide is used for animal feed' (ibid), with the USA the largest producer, followed by Brazil. Intensive livestock farming for meat and dairy production illustrates the interconnectedness and complexity of the global food chain and the role of food production and consumption in greenhouse gas emissions.

Global production of meat and dairy is set to rise significantly 'over the coming decades' (FAO 2006, 275), leading to 'further conversion of natural habitats into cropland' and increasing 'livestock's contribution to anthropogenic greenhouse gas emissions' (276). Meat consumption rises in middle to high-income level consumers in the developed world, but increasingly now in developing countries. Many civil society organisations have called for a reduction in meat consumption in the developed world (FOA 2006; FoE UK 2008; Compassion in World Farming and Friends of the Earth 2009), with a move away from intensive farming to a more humane, equitable and sustainable system of food production. For Compassion in World Farming and Friends of the Earth UK, this would be based upon the adoption of a 'fair, less meat' diet, where meat consumption in 'North America, Western Europe and Oceania' decreases and increases in 'Northern Africa, Western Asia and Sub-Saharan Africa for nutritional reasons as current consumption of animal products is relatively very low' (Compassion in World Farming and Friends of the Earth 2009, 12).

The message from the UN's 2006 report into the livestock sector's contribution to global greenhouse gas emissions has been slow to translate into public communication campaigns. Indeed, the report comments that 'civil society seems to have an inadequate understanding of the scope of the problem' (FAO 2006, 282). This inadequacy has been partially addressed in the last couple of years, as a number of civil society campaigns to reduce or eliminate meat consumption as a means of addressing climate change have been launched across the globe. The highest profile campaigns have been those focused upon the reduction of meat consumption rather than elimination, mainly through the global 'meatless' or 'meat free' Monday campaign (Meatless Monday 2010). Campaigns focusing upon the elimination of meat (and to some extent, dairy) have tended to been adopted by already established animal rights organisations, such as PETA (People for the Ethical Treatment of Animals) (USA) and Animal Aid (UK), who have been more explicit about promoting vegetarianism and veganism – the elimination of meat and dairy products from a person's diet – to help mitigate climate change. Indeed, it is only very recently that one of the international environmental NGO's, Friends of the

Earth UK, has produced a specific climate campaign targeting the meat and dairy industry, called The Food Chain Campaign.

The Food Chain Campaign focuses upon changing production processes rather than consumer practices. Its explicit claim that it is 'not about being vegetarian' (FoE UK 2009) reveals some of the problems involved in campaigns aimed at changing people's food consumption more broadly, but meat and dairy consumption more specifically. Such problems point to the symbolic nature of food and its intimate linking with identity and social status, as this is experienced through the body (Lupton 1996). Meat, in particular, is associated with masculinity and higher-income, middle-class groups (Adams 1990; Lupton 1996; Allen and Baines 2002). As Allen and Baines point out '…red and white meat have come to symbolise masculinity, high SES [socio-economic status], strength, and human domination over nature' (2002, 119). At the same time, however, the discourse of healthism emergent in consumer culture of the 1980s, and promoted by western governments in the 1990s, has categorised red meat as unhealthy. The purportedly high fat diets of lower socio-economic groups have been the target of healthy eating campaigns, in which a discourse of the neoliberal consumerist subject is deployed to assign blame to individuals, rather than addressing the social and economic structures which guide and constrain our food 'choices'. Attempts to change people's eating practices thus generate a diverse, contested and changing set of meanings.

Consumer research by Allen and Baines shows that people's attitudes to meat consumption are influenced by an individual's set of values or 'self-concepts', where the symbolic meaning of meat is evaluated 'against the values that they endorse' (Allen and Baines 2002, 127, 119). When meat was associated with 'dominance and hierarchy within human-to-human relationships' (ibid, 120), consumers in the focus group who did not identify with those particular values subsequently altered their dietary choices, consuming less meat. Campaigns aimed at changing people's levels of meat and dairy consumption as a means of addressing climate change, must navigate existing symbolic meanings of meat and dairy in order to change them. As Carol Adams has argued, the very concept of meat is reliant upon living animals becoming 'the absent referent', because 'animals in name and body are made absent for meat to exist…If animals are alive, they cannot be meat' (Adams 1990, 40). To what extent can the concept of meat as a commodity – reliant upon the commodification of nature and the maintenance of reductionist human/nature distinctions (Plumwood 2002) – be questioned through the linking of climate change to meat and dairy production/consumption?

Changing individual consumer behaviour through education – Meat Free Monday campaign

The global Meatless/Meat free Monday campaign specifically links meat consumption to climate change and is targeted at changing the consumption practices of individuals, communities, schools, businesses, public institutions and governments. Originally launched in 2003 as a public health awareness campaign by the John Hopkins School of Public Health (USA), Meatless Monday focused upon the human and environmental health benefits of reducing meat consumption for one day of the week (Meatless Monday 2010). In the last couple of years, nine other countries have supported the meat less or meat free Monday message, including the UK, Canada, Taiwan, Brazil, Holland, Croatia, Norway, Japan and Australia (ibid). The focus upon reducing meat consumption rather than promoting vegetarianism marks the campaign out as a 'small and painless' step (WWF-UK 2008a, 5), which does not promote bigger changes to lifestyle or consumption practices. By quantifying changes to consumption practices for one day a week, the climate change message is made tangible through calendar/clock time, realised through consumer practices. It places responsibility on the individual, and as such, puts the site of the consumer at the centre of climate politics. Conversely, however, the terms 'meatless' and 'meat free' linguistically and discursively work upon the assumption that meat is a normative constituent of meals (Heinz and Lee 1998), and as such, do not critique the commodification of animals as food.

Receiving a high level of UK media coverage, Meat Free Monday (MFM) was launched in the UK on 15 June 2009 by the singer-songwriter and ex-Beatles member, Paul McCartney, and his two daughters, Stella (fashion designer) and Mary (celebrity/fashion photographer). As high profile celebrities in the creative industries, their fronting of the campaign represents 'celebrity environmentalism' (Brockington 2008, 553). However, as instigators of the campaign, and as lifelong vegetarians and committed animal activists, they provide a more credible ownership of the campaign than simply lending their names as an endorsement, showing their difference from celebrities who are associated with less controversial conservationist projects (ibid). As patron of The Vegetarian Society since 1995, as well as campaigning for PETA, Paul McCartney has a long career in animal rights activism. His late wife Linda established the Linda McCartney vegetarian foods brand in 1994; a legacy and business maintained by Paul since her death in 1998 (Paul McCartney 2010). Fashion designer Stella McCartney does not use any leather or fur in her clothing collections, includes environmental 'tips' on her website, and funds a range of environmental and animal rights charities (Stella McCartney 2010), while photographer Mary McCartney has undertaken photographic shoots for PETA (Mary McCartney 2010). Established in their respective fields of music, fashion and photography, their credibility as creative practitioners, as well as celebrities (particularly Paul and Stella), does, however, bring a set of significations to the campaign that draw upon their cultural capital in the field of entertainment and as animal rights activists.

Their professional credibility is capitalised upon when they come together as the collective McCartney family in the campaign. Presented as lifelong vegetarians, the family narrative is visualised on the dedicated website for the MFM campaign, featuring a family portrait-style photograph of Paul, Stella and Mary, accompanied by the MFM logo. The logo is a reworking of the original brand logo for Linda McCartney's meat free food range, using the heart shaped smiley face from the food brand logo as the Meat Free Monday logo. The McCartney family brand is thus established through visual signifiers which figure the family as credible and knowledgeable representatives of the campaign by virtue of their long term involvement with animal rights and vegetarianism.

The website is a key information and communication tool for the campaign. The relative lack of imagery and large amount of text included indicates that the campaign is interested in presenting the information about the contribution of meat consumption to climate change, rather than through campaign slogans or imagery. The McCartneys' vegetarian activism is downplayed as the MFM campaign is characterised as 'an environmental campaign to raise awareness of the climate-changing impact of meat production and consumption' (MFM 2010a). The viewer is addressed positively as someone who 'cares about what you eat and about what's happening to our climate' (ibid), thus linking food consumption to climate change. A discourse of health positions the individual as a predominantly middle-class consumer who cares about 'what you eat' (ibid). With no reference to vegetarianism on the home page, meat production and consumption is framed discursively through climate change: the individual is encouraged to 'pledge to do your bit' (ibid), downplaying discourses of animal rights (Singer 1995) and emotion, usually 'central to the vegetarian position' (Lupton 1996, 124). The commodification of animals involved in the production of meat as food is not questioned, and the 'basic assumption that cooking with meat is the norm' (Heinz and Lee 1998, 91) is conversely reinforced through the identification of one day a week as being meat free.

This 'meat free', rather than vegetarian message (which would more radically question the social norms of meat consumption) is reinforced through the meat free recipes offered on the website by Linda McCartney, the Vegetarian Society and a range of celebrity chefs from top British restaurants. Radical changes to existing lifestyles are not advocated. Instead, giving up meat for a day is offered as an easy option that purportedly fits in with existing lifestyles choices. Recipes produced by Linda McCartney offer basic foods like shepherd's pie, pizza and meatless bolognaise sauce. Accessing these recipes redirects the viewer to the Linda McCartney Food website, which also acts as an advertisement for her ready-made vegetarian frozen foods; implying that Linda's meat free recipes can also be replaced by her ready-made products. The celebrity chefs offer more luxurious and complex recipes: 'beetroot ravioli with cashew-chive boursin cheese' from Chad Sarno, chef for Saf vegetarian and vegan restaurant in London; or 'ricotta stuffed courgette flower zucchini salad' from Francesco Mazzei from L'Anima Italian restaurant in London. The 'simple and painless steps' discourse encouraged by the campaign is cut through with a middle-class, aspirational discourse, where the celebritisation of restaurant chefs, and the types of foods they use in their recipes, appeal to middle-class tastes. At the same time, the recipes function as advertisements for the restaurants (many of which are not vegetarian).

The viewer is addressed as a consumer citizen through the information given under the main heading 'How can I help the planet?'. Presented in descending order as 'less pollution', 'making a political statement', 'alleviating world hunger', 'better health' and 'protecting animal rights', the reduction of meat consumption draws upon discourses of environment, politics, health, human ethics and animal welfare to produce an ethics of care through consumption. Consumer action is presented as a form of politicised citizenship, which takes emphasis away from policy changes:

> Politicians follow. They don't lead. And because of meat's association with affluence and the fear that asking people to eat less meat might make them unpopular, most politicians shy away from this issue… So the best hope for change lies in average people becoming more aware of the true costs of industrial meat production and taking action themselves (MFM 2010b).

Each piece of information presents a rational overview of the impact of meat consumption on the climate, human health and, to a lesser extent, animal welfare. The rationalisation of these reasons positions the viewer as an individual consumer able to make appropriate choices through information; criticised by many as failing to take into account the symbolic dimensions of people's consumption practices (Hobson 2006; Clarke 2008). An appeal to the self is made without referencing or activating more politicised identity categories such as vegetarian or vegan. Overall, reducing meat consumption is presented as an ethically informed rational choice, which individuals can sign up to supporting, and can donate money, following reasoned consideration of this environmental issue.

The linking of meat consumption to climate change decreases the time-space distanciation between commodity meat production and its climate impact, by linking food production and everyday consumption to greenhouse gas emissions. Through a positive appeal to the efficacy of individual action, everyday actions are linked to global effects. Yet, at the same time, the relationship between food consumption and climate change is abstracted by making an environmentalist appeal to 'the planet': 'By making a simple change in the way you eat, you are taking part in a world changing campaign where what's good for you is also good for the planet' (MFM 2010b). Although the linking of the self to the planet positively localises responses to mitigate global climate change, the consumer actions taken in the present are figured as part of an abstract future: 'The scale of the problems we face can make many of us feel helpless, and yet each of us has the power to make changes in our lives that can have a meaningful impact on the future' (ibid: Help the planet page). Climate change is rendered a distant event, even as it is linked to everyday food practices.

The campaign makes an important contribution to the climate message by linking meat production and consumption to climate change, as well as mainstreaming climate action through the promotion of one 'meat free' day of the week. The quantification of meat consumption in relation to calendar time makes the individual action meaningful to a potentially large number of people, because eliminating meat one day a week is linked to the reduction of global greenhouse gas emissions. Using the slogan 'Support Meat Free Monday', the campaign employs positive rhetoric. The campaign, however, also reproduces an approach criticised for assuming a rational individual who changes behaviour as a result of information, leaving (middle-class) lifestyle practices unchanged. Furthermore, while making visible the role of meat consumption in the production of greenhouse gases works to foreground human actions as part of the environment, the failure to question the concept of meat as a commodity reinforces human/nature distinctions through the commodification of animals for human consumption.

To what extent does the campaign question consumerism itself? The McCartney family represent a brand whose name is not only used to promote highly processed and freezer-dependent (thus highly energy profligate) vegetarian food through the Linda McCartney frozen food range, but whose occupations promote an aspirational consumerist lifestyle through the production of luxury goods by Stella McCartney and the celebrity/fashion photography of Mary McCartney. The McCartney brand and its links to celebrity are reinforced through the list of celebrity names and celebrity chefs (and their restaurants) who support the campaign via a presence on the website. Although kept to a minimum, photographs of the McCartneys are present in the form of a family portrait on the homepage, a photograph of Paul McCartney standing in a field of organic spelt which he owns, and a snapshot of Linda McCartney on the recipe page. Furthermore, the use of quantitative clock/calendar time to help change consumer actions, alongside the discourse of environment which partially positions 'the planet' as separate from humans, downplays the urgency of addressing climate change through major lifestyle changes, rendering climate change a future event, rather than a present reality.

As one of only a few high profile public campaigns linking meat to climate change, MFM does represent an important start to the mainstreaming of climate action through meat (although not dairy) consumption. Furthermore, it enables the message to be promoted by other organisations and groups. The UK animal rights organisation, Animal Aid, is working with local councils to make schools commit to a Meat Free Monday, simultaneously promoting a more explicit vegetarian and vegan message (Animal Aid 2010). PETA has been promoting Meat Free Monday in primary and secondary schools in the UK, linked in to promoting an animal welfare and vegan message.[2] Globally, the 'meat free' Monday campaign has been adopted by numerous schools and cities, including Bremen in Germany and San Francisco in the USA (ibid). While the UK MFM campaign may downplay more radical lifestyle changes, such as the adoption of a vegetarian or vegan diet, in place of a more mainstream individualist appeal, its message has the potential to transform community and institutional practices through the networks of campaigning groups it facilitates.

2 My thanks to Suzanne Barnard, Education Manager, at PETA foundation UK for providing this information.

Activating an ethical individualist self – Peta2's Meat's Not Green campaign

Launched online in August 2008, Meat's Not Green is a campaign by Peta2, the youth division of PETA (USA). Initially, its aim was to petition the USA Food and Drug Administration for legislation that all meat products have a warning label stating that meat is the number one cause of global warming. This aim was eventually changed to encourage individuals to pledge to go vegan for 7 days.[3] Like MFM, Meat's Not Green explicitly links meat consumption to climate change (although the term global warming is used in the campaign), but includes the role of dairy consumption in the production of greenhouse gases. In contrast to MFM, this campaign aims to eliminate meat and dairy produce from people's diets, rather than simply reducing meat consumption. The slogan, 'Meat's Not Green' is a negative proposition that assertively draws upon the discourses of animal rights and environmentalism to present its message. These discourses are made more explicit in the tagline – 'Save the planet – Go vegetarian', reinforced by the slogan's logo. Here, the image of a calf between the words 'Not' and 'Green' visually reinforces the animal rights message, while the inclusion of a recycling logo in the 'o' of 'Not' utilises a sign from environmentalist discourse. The inclusion of the calf acts as a sign for the meat and dairy industry, whose production processes are based upon the removal of calves from dairy cows for the veal industry, as well as to access the cow's milk for human consumption. The combination of signs indicates an audience more familiar with animal rights/welfare issues than environmentalism, given the incongruous inclusion of a recycling sign to signify the environment.

With significantly less text than the MFM website, this website is more closely linked to the discourses and rhetorical strategies of animal rights and environmental activism, with a campaigning style tone deployed: 'What's the number one cause of global warming? Cars? Nope. Factories? Nope. Planes? Nope. According to the United Nations, raising animals for food generates more greenhouse-gas emissions than all the cars, planes, ships, trucks, and trains in the world combined' (Peta2 2010). As an existing animal rights organisation, the pledge to go vegan for seven days is potentially aimed at existing vegetarians, explicitly incorporating the discourses of animal rights with those of the environmentalism: 'I want to help save the planet and to stop supporting cruelty to chickens, pigs, cows, and other animals who are raised for food. By signing my name, I pledge to try going vegan for a week' (ibid). Although it uses days as a means of quantifying behavioural changes into manageable actions, unlike MFM the emphasis here is upon a more explicit assertion of the self, where the 'I' activates an identity politics that is reflected in the pledge itself, to go vegan. The act of giving up meat and dairy products is named as an identity, placing this kind of pledge within the identity politics of animal rights. Furthermore, to pledge means to solemnly promise, whereas

3 My thanks to Abi Izzard at Peta2 for supplying me with information about the campaign via email.

to support (as in MFM) connotes a less involved form of behavioural change that does not alter, or call into question, a sense of self. Identity politics are reinforced further through the activities offered on the website which include creating your own slogan to go on a warning label for meat (the original aim of the campaign) and posting the individually sloganed image up on your Facebook or MySpace page. An application, in the shape of an ipod, that calculates how green your diet is, can also be copied to your Facebook and MySpace pages, thus linking the activist self to a community through social media.

In the promotional campaign video, dominant (and normative) visual tropes of climate change – such as satellite images of planet earth and hurricanes, as well as images of droughts, melting glaciers and flooding – are reproduced and recontextualised in discursive relation to animal activism to promote a new discourse of climate change in relation to meat and dairy consumption. Images of climate impacts to visible landscapes are accompanied by the female voiceover who states, 'We know you're concerned about the planet and its inhabitants. You wanna help stop global warming. But what can you do to make a difference besides changing some lightbulbs?' (ibid). Existing visual discourses of climate impacts to a visible nature help give meaning to a new climate discourse. Yet existing discourses of climate mitigation are undermined through the posing of rhetorical questions that humorously devalue the efficacy of certain individualist behavioural changes, such as changing light bulbs. The voiceover asks us, 'What really is the number one cause of global warming?' (ibid): Failure to recycle, and the use of cars and planes are the different visual suggestions, each of which is followed by a cross imposed upon the image to signify 'no'. An image of a piece of steak on a plate then appears, overlaid by the spoken question, 'What about that steak on your plate?' The voiceover then reiterates the written message of the pledge. 'According to the United Nations, raising animals for food causes more greenhouse gas emissions than all the cars, planes, ships and trains in the world combined' (ibid). The voiceover is visually reinforced though a set of images associated with animal rights and climate change: video images of livestock crammed together and satellite images of hurricanes, flood water and a polar bear.

The choice of animal images showing suffering encountered in factory farming makes a visual and discursive link between animal cruelty – thus revealing 'the absent referent' (Adams 1990, 40) – and the causes of climate change. Throughout the video, spoken references to animals are visually reinforced through images of animals suffering in cramped living conditions. As such, two messages are communicated simultaneously through visual/verbal signs: that the production of animals for human consumption contributes to climate change through greenhouse gas emissions involved in rearing animals and in the destruction of rainforests for meat production; and that the production of animals as commodities for human consumption is cruel and inhumane. The absent referent in meat and dairy production is revealed through the images of animal suffering; familiar strategies for the campaigning tactics of animal rights organisations such as PETA, but used here to also make visible the contribution of meat and dairy production to climate change.

The greenhouse gas emissions of everyday practices in the form of meat, dairy and egg consumption are made tangible and explicit through spoken and visual text. Yet, the images of climate change used to convey impacts to uninhabited land reproduce dominant visual imagery of climate change as separate from humans through reference to an objectified visible nature. Instead, humans are included in the video as both perpetrators of destruction (involved in animal rearing and as consumers of meat/dairy) and as agents of individual change. 'You can help' is accompanied by a speeded up video montage of factory farm scenes, communicating the message that eating meat is cruel. The individual is then asked to sign Peta2's petition 'for all meat packages to be labelled with a clear warning that meat is the number one cause of global warming' (ibid). While the petition is aimed at achieving policy changes through the USA Food and Drug Administration, positioning the individual as a citizen, the video ends with the assertion that 'the consumer has the right to know', thus utilising consumerist discourses. The campaign both questions and reinforces nature/culture and human/animal distinctions in its efforts to make visible the role of industrial meat and dairy production/consumption in the creation of climatically changing greenhouse gas emissions. While the campaign importantly reveals this relationship through commodity meat/dairy production and consumption, thus critiquing the commodification of animals and environment, it does so by visually and textually presenting the animals and a climatically changing environment as separate from humans, even as it extends the right to life, to animals.

Compelling in its affirmation of action, the individualist discourse of the climate actions proposed – 'I pledge to go vegan' – is indicative of what Plumwood calls 'excessive individualism' in the context of an animal rights framework (Plumwood 2002, 152). Excessive individualism turns attention,

> away from the structural origins of the atrocities committed against animals in the factory farm and the commodity framework to the question of the virtue of individual consumers who makes a choice of eating animal food, no matter how obtained (ibid, 153).

For Plumwood, this approach limits the integration of 'animal justice concerns with those of environmental and human justice movements' (ibid). I agree with Plumwood that justice movements need to involve animals, environment and humans as unseparated, and therefore I recognise the individualist limitations of the campaign. However, I also want to acknowledge the important contribution of Peta2's campaign to a developing discourse of climate change by making visible the link between industrial meat and dairy production/consumption and global greenhouse gas emissions, and offering more radical behavioural changes than MFM. The question of the structural origins of factory farming, and its relationship to climate change, not focused upon in this campaign, is specifically addressed by FoE UK's The Food Chain Campaign.

Regulating unsustainable food farming practices – Friends of the Earth UK's The Food Chain Campaign

In contrast to both the MFM and Meat's Not Green campaigns, FoE UK's The Food Chain Campaign does not encourage individual behavioural change through reducing or eliminating meat and dairy products from a person's diet. Instead, the campaign focuses upon the production processes involved in the meat and dairy industries, mainly the industrial production of soy used for animal feed in factory farming and the role of the UK Government in supporting these unsustainable food practices. Like the MFM and Meat's Not Green campaigns, the link between soy production for animal feed and climate change is foregrounded: 'Vast areas of land have to be cleared to grow the soy, causing: deforestation; greenhouse gas emissions; [and] the loss of valuable wildlife habit' (FoE UK 2009). Rather than targeting the individual consumer to change their eating habits at the point of consumption, FoE focuses attention upon UK Government subsidies for intensive factory farming, which, it argues, supports the large-scale production of soy, and subsequent deforestation in South America. The aim of the campaign is to divert government subsidies away from unsustainable farming and to encourage local, 'planet-friendly farming' (ibid). As such, the climate impacts of intensive factory farming are communicated through highlighting the economic unsustainability of these practices, with policy regulation of livestock farming the key message of the campaign.

The campaign reveals the set of commodity chains involved in meat and dairy production, and the environmental, social and economic consequences of large scale, intensive factory farming. While the MFM and Meat's Not Green campaigns also foreground the climate impacts of land clearance for soy production used in livestock rearing, FoE's campaign is more explicitly focused upon changing these processes rather than eliminating or reducing the final commodity in the form of meat and dairy produce. Fronted by FoE's Senior Food Campaigner Clare Oxborrow, the campaign video, 'Fix the Food Chain', begins by presenting the economic case against factory farming: 'At the moment £700 million pounds of taxpayer's money is spent propping up factory faming of livestock in England and billions of public money is invested every year in providing meat and dairy for schools, hospitals and care homes without a thought of how it is produced' (FoE UK 2009). The economic profligacy of factory farming is then linked to the environmental and social impacts of soy production for animal feed in South America, which involves destroying rainforests and 'forcing out communities off their land' (ibid). The social impacts of soy production upon communities is highlighted by neither of the other two campaigns, although they all use the same statistics from the UN; that 'meat and dairy production causes 18 percent of the world's climate-changing emissions. More than all of the planes, cars and lorries on the planet' (ibid).

The video images accompanying the audio messages function as visual evidence of the processes involved in meat and dairy production and those who are affected: land, humans and animals. FoE's campaign seeks to link animal, environmental and human justice issues together. Significantly there are no images of climate impact to land or humans; instead there are images of cleared (deforested) swathes of land, a group of South American people sitting in a communal area and going about their daily routines, such as washing clothes in a river, South Americans marching in protest, and a series of images of animals in factory farms. These images function as a way of bringing a broader set of understandings about the contribution of meat and dairy production to climate change, and the economic and social impacts of livestock rearing on communities, foregrounding issues of social and, to a lesser extent, animal justice. Interjecting the video images and spoken text are a series of cartoon cut-outs of a chicken, pig, cow, sack of soy feed, chainsaw and logs, and a balloon with CO2 written on it. Each image functions as a representative and visible sign within the chain of commodity food production, which includes the effects of meat and dairy production through the creation of carbon dioxide as a result of deforestation and from the growing of soy for animal feed. In doing so, the images function to defetishise the commodity production of meat and dairy by calling attention to some of the processes involved in these industries, which are also dependent upon other commodities, such as soy and land, to produce the final products. Using similar style cut-outs, other images (see Figure 4.1) are less visually explicit about the links between the commodity carbon chain, instead communicating that animal farming should be free-range to ensure 'planet-friendly farming' (FoE UK 2009).

Figure 6.1 The Food Chain Campaign from Friends of the Earth UK

Source: Friends of the Earth UK, 2010.

What makes FoE's campaign different from the previous two is that it focuses upon government regulation, rather than individualised consumer action, assigning responsibility to the economic and policy structures of the UK Government which support the global meat and dairy industry through financial subsidies for intensive livestock farming. The wider regulatory system involved in food production is the focus of attention, rather than consumption of the end product, in this global food chain. Activism does not involve 'getting people to go vegetarian, though it won't hurt to eat less and better meat and dairy' (ibid). Instead, the consumer is framed as a citizen, urged 'to put pressure on politicians to change the system and change the deal behind our meals' (ibid). Here, the action moves away from individual choices about consumption of meat and dairy, to the wider economic and regulatory systems which support intensive and, by inference, inhumane farming. Rather than an individualised campaign which utilises identity politics to promote the reduction or elimination of meat and dairy consumption, FoE's campaign explicitly distances itself from such individualised strategies, instead targeting the UK Government to 'support small-scale farming and develop home-grown animal feed' (ibid). Thus, 'The Food Chain Campaign isn't about getting rid of meat and dairy farming – it's about getting government to revolutionise it' (ibid). In downplaying the role of the consumer, the campaign importantly draws attention to the wider systems which support intensive farming, identifying the individual as a rational citizen who 'is the strongest link in the food chain' (ibid) by virtue of their political voice, while drawing attention to the global commodity chains involved in meat and dairy production. While global meat and dairy production is defetishised through the identification of commodity carbon relations, the social norms of meat (Heinz and Lee 1998) and dairy consumption are left unaddressed. As such, animals remain commodified as food products.

Discussion – negotiating appeals to individualist action and policy regulation

Reproducing the statistics produced by the UN's 2006 report into the livestock sector, all three campaigns explicitly link meat (and to a lesser extent dairy) consumption to climate change, revealing the role of soy production for global animal feed which involves the clearance of vast swathes of forest and land for crops and for livestock rearing. Yet where MFM and Meat's Not Green focus upon the role of consumption in the commodity food chain, targeting the individual as the site of consumer action, FoE UK foreground the production processes in livestock farming to argue for localised and small scale farming achievable through political action via pressure on politicians. Of the three campaigns, Peta2's Meat's Not Green campaign explicitly appeals to an identity politics, addressing the individual as a vegetarian or vegan through an emotional appeal to animal rights to communicate the climate message. In contrast, MFM avoids reference to vegetarianism or veganism, to offer a rationalised approach to reducing meat consumption through an environmentalist and ethical appeal to the well-being of the individual and 'the planet'. Avoiding identity categories in its rhetorical appeal, MFM dismisses the role of politics in addressing climate change, instead advocating a politicised consumer position, who in reducing meat consumption, also alters consumer practices. FoE's Food Chain Campaign similarly avoids identity categories, but in contrast to the other campaigns does not explicitly advocate the reduction or elimination of meat and dairy products, placing emphasis upon the economic, environmental and social unsustainability of global meat and dairy production in order to advocate local farming, as a more sustainable, and thus rational, development.

To some extent, all three campaigns rely upon existing discourses of climate change to reframe meat and dairy production and consumption as contributors to global greenhouse gas emissions. MFM assumes a more knowledgeable (middle-class) audience, utilising an environmental discourse of climate change as affecting 'the planet', whilst also drawing upon discourses of health to position the viewer as someone who cares about the health of themselves and the environment. Meat's Not Green deploys video images of climate impacts to land (devoid of humans) alongside those of animal suffering, to establish a point of emotional equivalence between the two. While figuring the individual as able to contribute to climate change mitigation through the elimination of meat and dairy from their diet, the discourse of animal rights alongside the images of climate impacts work concurrently to remove the human from the environment. An environmental justice organisation, FoE UK's long history of campaigning for climate change is illustrated in its deployment of discourses which may not be so readily associated with climate change, foregrounding an economic and sustainability discourse that highlights the complex role of food production in the politics of climate change.

In defetishising the commodity relations involved in meat and dairy production, all three campaigns offer significant contributions to cultural understanding of climate change, by making it meaningful at the level of everyday food consumption. For any changes to consumer behaviours to take place, the social and symbolic relevance of consumption practices needs to be acknowledged. In this context, MFM and Meat's Not Green offer the more explicit engagement with consumer habits by advocating changes to consumer practices that can be understood in the context of a weekly calendar, while Meat's Not Green also reveals the suffering involved in factory farming through photographic and video images. However, this also constitutes the limitations of these campaigns. For MFM, the simple and painless steps approach (which include accessing recipes on the website) does not really question an existing sense of self, thus it may not translate into more radical behavioural changes, particularly in relation to a middle-class consumerist lifestyle. For Meat's Not Green, there is no sense of understanding why people may eat meat or dairy, despite being aware of the cruelty issue. For FoE's The Food Chain Campaign, there is no discussion of consumer habits, nor of the symbolic nature of food, because it does not advocate changes to consumer behaviour but rather policy regulation. Promoting an ethics of sustainable consumption, however, involves changing people's consumer practices through education and moral/ethical appeals, as well as providing government laws and incentives (Gardner and

Stern 1996, 27 cited in Jackson 2004, 1040). Taken together, the three campaigns activate all of these strategies.

Conclusion – towards an ethics of sustainable and just consumption

As consumerist cultures increase across the globe (WorldWatch Institute 2010), the consumption of foods, goods, services and energy rises, putting further pressure on the planet's resources, already unsustainable at European and North American rates of consumption (WWF International 2008; nef 2009a). The energy and carbon intensive practices of consumerist societies, and the commodity chains which mask these processes, thus need to be made more visible, and actively reduced, by individuals, societies, businesses and governments. This chapter has argued that consumerism, as a way of life, needs to be critiqued (Gilbert 2008), and that those of us living in consumer societies need to consume both less and differently. This does not entail simply buying 'green' or 'ethical' products in place of others, but rather being cognisant of the energy required for the production, transportation, consumption and disposal of all goods, as well as services, and changing our consumer habits/practices. Yet, information about the carbon intensity of consumption practices is not sufficient to alter people's behaviour (Hobson 2006; Clarke 2008). We need also to be attentive to the symbolic and cultural meanings associated with particular commodities and acts of consumption, which contribute to our sense of self and social status. In doing so, the symbolic saliency – or fetishisation (Marx [1867] 1990) – of certain commodities/consumption practices can be changed (Heinz and Lee 1998; Allen and Baines 2002). At the same time, acts of consumption need to be understood in the context of the structural systems which support and constrain consumer practices: activating sustainable consumption requires government legislation.

This chapter has analysed three public communication campaigns which seek to make visible the role of meat and dairy industries in the production of greenhouse gas emissions and resultant climate change. One of the aims of this book has been to argue that climate change needs to be made culturally meaningful and relevant to people's everyday lives. The linking of food consumption to climate change is an important way of achieving this. Peta2's Meat's Not Green campaign utilises established moral discourses of animal rights to set up an equivalent discourse of a climatically destroyed environment as a result of meat and dairy consumption. It focuses upon individual action at the point of consumption, advocating a vegetarian or vegan position through the explicit activation of identity consumerist politics. In contrast, MFM does not utilise identity politics in the form of a vegetarian or vegan identity, instead promoting 'small and painless steps' to reducing meat (but not dairy) consumption one day of the week. It links both meat production and consumption to a wide range of issues, including health and animal rights, but more explicitly utilises an environmentalist discourse to encourage the reduction of meat consumption. Unlike Meat's Not Green, and FoE's campaign, it explicitly undermines the role of governments in changing consumption practices, articulating an individual consumerist position for action. Both MFM and Meat's Not Green, I would argue, endorse people's existing values, in order to promote a meat free or 'meatless' position. Meat's Not Green utilises an animal rights and identity politics framework to encourage the elimination of meat and dairy consumption, while MFM utilises an environmental discourse, familiar to a middle-class audience, to encourage the reduction of meat consumption.

In contrast to these campaigns, FoE UK's The Food Chain Campaign focuses upon changing government policies which economically subsidise industrial factory farming. As such, the commodification of animals as food products is not viewed as problematic, but rather the economically and ecologically unsustainable practices of factory farming. Individuals are targeted as citizens who should campaign to change government laws to promote local and sustainable farming, rather than activating change at the point of consumption through the elimination or reduction of meat and dairy consumption. The Food Chain Campaign more explicitly reveals the globalised and unjust nature of industrialised commodity meat and dairy production, to include the unjust treatment of animals, humans and land, as part of the environment. Of the three campaigns, FoE's provides a more cohesive overview of the commodity chains involved in the factory farming of meat and dairy, including the social and environmental effects of its production. As such, FoE more explicitly attempts to link an ethics and sustainability agenda of animals, humans and environment as interconnected and unseparated (Plumwood 2002).

It could be argued that FoE's campaign is more successful in revealing the relational links between meat and dairy production and climate change, in presenting a more complex understanding of global meat/dairy production as unsustainable and unjust, and for advocating more structural, rather than individual changes. Yet, in choosing to ignore meat and dairy consumption, the campaign, I would argue, continues to figure animals as consumer products, failing to question the commodification of nature (Plumwood 2002, Adam 1998a). The campaign employs a rationalist approach which denies any emotional engagement with these issues, and in doing so, ignores an important dimension of people's capacity to engage with climate change. Any campaign which attempts to change people's diets is always at risk of being accused of preaching, by questioning a person's 'individual right' to consume what they like; particularly for commodities like meat and dairy that are regarded as both natural and aspirational in western consumer societies (Heinz and Lee 1998). This should not lead, however, to a rationalist – or emotionally disengaged – approach to understanding the role of particular forms of consumption in contributing to climate change, as this devalues affective responses to climate change, reinforcing binary distinctions which present humans as rational and masculine, and nature as feminine and irrational (see Chapter 1 for a more in-depth discussion of these distinctions). An ethics of sustainable consumption must appeal to our emotions, our ethics, our values and our rational selves in an attempt to change social norms about consumption.

While all three campaigns analysed offer important contributions to the communication and making meaningful of climate change in our everyday lives, I would argue that the commodification of animals necessary for the production of meat and dairy products – 'the absent referent' (Adams 1990, 40) – needs to be made more visible and explicitly critiqued in the context of climate change. Firstly, such a process of commodification reinforces nature/culture distinctions, which have contributed to a separation of humans from the environment (and thus an understanding of climate change); and secondly, in the context of energy-intensive consumer societies, such commodification erases the processes involved in commodity production which imbue products with symbolic meaning. In the context of climate change, the contribution of industrial meat and dairy production to global greenhouse gas emissions; the unjust treatment of animals, humans and environment within these industries; and the reproduction of commodifying processes as part of a consumerist culture, means that meat and dairy consumption within western societies must be radically reduced or eliminated. In seeking an ethics of sustainable consumption, I agree with Plumwood's call for a 'critical, contextual and ecological form of vegetarianism that is …inclusionary' (Plumwood 2002, 154) rather than exclusionary. I would go further than Plumwood to advocate a vegan diet, given that dairy cows are a significant component in the global livestock sector. Within western societies, the choice to not eat meat or dairy represents both a reduction in particular forms of consumption as well as a decision to consume differently. As part of an ethics of sustainable consumption, a vegan position, however, must acknowledge the contextual nature of such a 'choice' within a western context, and must be supported by government legislation and incentives which promote sustainable consumption, where an ethics of human, animal and environmental justice are intertwined.

Chapter 7

Imaginative Engagements: Critical Reflections on Visual Arts and Climate Change

> We can register what is happening with satellites and scientific instruments, but can we register it in our imaginations, the most sensitive of all devices? (McKibben 2005).

Since Bill McKibben made his appeal to artists to help make climate change culturally meaningful through creative practice, arts engagement with climate change has grown considerably. Large scale projects like Cape Farewell (UK), which brings together artists, scientists and communicators to 'stimulate the production of art founded in scientific research' (Cape Farewell 2010a), co-exist with the more explicitly activist work of organisations such as PLATFORM (UK), who utilise art to campaign on issues of ecological justice. The current surge of art and climate change projects was further demonstrated at the Copenhagen conference, where the National Gallery of Denmark's visual arts exhibition, *RETHINK: Contemporary Art and Climate Change,* took place alongside a range of smaller climate arts projects/installations by individuals, civil society groups and NGOs. In the UK, the Royal Academy of Arts also staged its *Earth: Art of a changing world* exhibition in December 2009 to coincide with the conference. Bringing together existing artwork not originally inspired by climate change, alongside new artwork specifically commissioned to address this issue, the exhibition illustrated the ways in which art can intentionally engage with climate change, or be actively reinterpreted through the lens of climate change. As such, cultural and creative engagements with climate change do not fix meaning, but present the possibility for multiple ways of perceiving, engaging and understanding.

Historically, the visual communication of climate change in popular media culture has tended to focus upon, and prioritise, photographic and film documentation of climate impacts, which have been criticised for their disempowering and disengaging effects (Doyle 2007; O'Neill and Nicholson-Cole 2009). Existing visual icons of climate change – the polar bear, melted glaciers, flooded urban areas, polar ice – have become so ubiquitous that they do little to foster a more active or embodied engagement with this issue. Furthermore, western culture's privileging of vision as a primary sense constructs (scientific) knowledge as objective and rational, through which the environment gets figured as a visible nature to be 'seen' and henceforth understood (Adam 1988a; Plumwood 2002; Macnaghten and Urry 1998). Such figurations can lead to sense of disconnect from our environment(s) and from the issue of climate change, which is rendered distant and future, rather than an immediate present. How can visual arts offer different ways of imag(in)ing climate change that activate different senses and experiences other than the visual?

This chapter explores the possibilities that visual arts offers for new ways of perceiving and understanding climate change, through a consideration of the ways in which visual arts engagement with climate change can appeal to our emotions, imaginations and our senses. As the interrelated concepts of nature, vision and time in western constructions of (scientific) knowledge have posed difficulties for making climate change both intelligible and culturally/personally meaningful, this chapter asks: to what extent can visual arts help us to (re)envisage both our existing and future social realities, and facilitate a more embodied and immediate response to climate change? Given the personal, cultural and contextual nature of an individual's responses to, and engagement with, art, I therefore explore the role of visual arts in the making meaningful of climate change through a series of my own critical reflections upon a selection of visual arts examples. These are taken from my experiences of visiting the *Earth* exhibition at the Royal Academy of Arts (London, UK), and Cape Farewell's *Unfold* exhibition (London, UK).

The *Earth* exhibition (3 December 2009 – 31 January 2010) at London's Royal Academy of Arts, UK, was organised by the Royal Academy in collaboration with Cape Farewell. Involving 35 international artists, the exhibition was timed to coincide with the Copenhagen conference, and brought together artists dealing explicitly with climate change (such as those involved with Cape Farewell), as well existing art work not originally conceived as a response to climate change. Established in 2001 by the artist David Buckland, Cape Farewell seeks to activate cultural responses to climate change by bringing together artists, writers, scientists, educators and media 'for a series of expeditions into the wild and challenging High Arctic' (Cape Farewell 2010b). This was expanded in 2009 to include an expedition to the Peruvian Andes. Alongside the production of art, literature and music, Cape Farewell produce educational resources for children/youths, as well as contribute to climate science through collaboration with scientists. During 2010, its touring exhibition, *Unfold*, brought together work produced by artists who have participated in an expedition to the Arctic (2007, 2008) or the Andes (2009).

Drawing on artworks included in the *Earth* and *Unfold* exhibition, this chapter is necessarily selective, and somewhat personal, offering reflections and raising questions through its exploration of the spaces of imagination and meaning opened up by visual arts. Alert, however, to the charges of cultural elitism often levelled at the arts, I reflect critically upon the discursive/institutional context of the space of the artwork's display and how this contributes to the possibilities and limitations of the making meaningful of climate change through creative and artistic practice. Through notions of belonging, place, space, time, vision, display and context, I both identify and explore how these themes offer ways for artists to help (re) imagine climate change as a present and pressing concern.

Belonging: remapping place/space

> Maps of the world will have to be redrawn, as global warming melts the Greenland ice cap, inundating coasts and major cities (Lean 2004).

I am standing at the doorway looking into a large room filled with thousands of small, but differently sized, terracotta clay figures standing upright. Packed closely together, the individual figures stare directly at me from a dimly lit space. With no distinguishing bodily features, such as arms or legs, it is the round shaped head on each figure, with two hollowed out eyes, which make them appear human. The figures spill out past the door, appearing to suddenly stop in a straight line, parallel to the wall outside, as if a barrier has been drawn across to hold them in. They appear like a mass of thousands of people confined to a space, desperate to leave. The shaft of light that falls across the room from the door creates the impression that these people have been locked into the darkness for a long time. Yet, there is no sense of panic, only a haunted sea of eyes, expectant but resigned to the wait, with nowhere to go. The effect is chilling. The stillness is broken by the persistent, but haunted gaze of the figures. I watch and am being watched. Are they climate refugees, displaced from their homes, their land, fleeing from environmental disaster, or from war over resources? Will I help them? How?

I look more closely and see that some figures stand taller than others, as the differences in each become more acute. These are individual people: friends, partners, mothers, fathers, children. I move my gaze out, and the space is redrawn, as their individuality is replaced by a vast collective, where slight differences in colour work on a mass scale to produce swathes of shades, clusters of which appear like land masses in the sea. I cannot look away, but want to: awed by the scale and horrified by the thought of what lies ahead for these people, for us, for me. They are a representation of humanity and inhumanity.

Anthony Gormley's *Amazonian Field* (1992) occupies an entire room at the Royal Academy's *Earth* exhibition (UK). Originally produced for the 1992 Earth Summit, the figures were created by local people from Porto Velho in Brazil, each one individually made. I saw a version of the field many years ago in Tate Liverpool, UK. Called *Field of the British Isles* (1993), the process of production was the same, each figure made by local people in St Helens, UK. I remember responding strongly to the work – being overwhelmed by a sense of humanity, vulnerability, intimacy and strength, but overall of human interconnectedness. Reconfigured in the context of climate change, the *Amazonian Field* appears more sinister, as the emerging social, political and geographical realities of a climatically changing world create a different viewing perspective and interpretation.

Occupying the gallery space outside of *Amazonian Field* is Mona Hatoum's *Hot Spot* (2006): a large tilted steel globe of average human height, with neon tube lights which demarcate the borders of the world's continents. The lights both hum and glow, as energy and movement is communicated visibly and sonically. The work invites a more conceptual contemplation than Gormley's. The mapped globe invokes a sense of political tension, as the glowing and humming boundary lines suggest conflict over these geopolitical borders. Maps seek to define, control and own space, yet the implication here is that these spaces are continuously contested. On one level then, the globe points to political unrest and conflict over the world's geographical spaces, and the ultimate futility of such endeavours to seek control. In the context of climate change, these boundaries take on different meanings, calling attention to both the literal geographical changes to place and space as a result of rising sea levels, and the political implications of these shifting shapes, as humans seek shelter in other places and spaces, necessarily crossing borders and national territories. The historical and ongoing conflicts over the world's geopolitical boundaries thus takes on a more disturbing perspective, as the reality of a climatically changing world and the creation of climate refugees puts further pressure on geographical spaces, already politically contested. The title, *Hot Spot,* simultaneously invokes a climatically hot place, as well as a politically contested space. Interpreted in relation to Gormley's work, the world's responses to climate refugees takes on a disconcerting guise.

Like Gormley, Hatoum's work was not originally intended as a commentary upon climate change. Yet, recontextualised, its original set of meanings take on a further dimension, as we reinterpret her work through a different set of knowledges and expectations. What makes both of them work as visual arts engagements with climate change is that they call upon the viewer's own set of values, knowledges and imagination to interpret their meaning. In dealing with our connections to place and space, and our potential dislocation from these, they ultimately ask us to think about ourselves as humans and our sense of belonging – to other individuals, communities, cultures, locales, nations and to the world. As a sense of belonging is crucial to the creation of identity, then these artworks alert us to the contingency of our sense of self and identity on our geographical surroundings and the socio-political networks within which we are embedded. Understanding these contingencies helps open up an emotional space for us to think about what we have to lose, and how we may harness these connections for positive climate action in our places/spaces of belonging, from which we may become displaced.

Making climate change meaningful means making it relevant to everyday lives and social practices. In this context, Cape Farewell's expeditions to the Arctic, and more recently the Andes, are problematic. Melting glaciers have figured prominently in media and NGO communication on climate change. This is not surprising given the unique sensitivity of the Arctic regions to climatic changes, as well as the visual metaphors that images of melting glaciers provide. Yet, these places remain both distant and over signified for western audiences. I am skeptical of the use of 'remote' places through which to engage people, particularly ones that are replete with geopolitical and colonial histories. The real challenge is making climate change culturally meaningful to people's own social realities, and sense of belonging. This does not necessarily mean simply reducing cultural responses to particular local spaces, because climate change is about making us question our sense of selves and our actions which may transcend the specificity of place.

Rather, it is about connecting to different places and spaces, attentive to the power relations that underpin these forms of belonging, and therefore also to those who are included and excluded from such spaces.

Immanent time: past/present/future

> Art dances a waltz with time; the three steps of past, present and future which can swing tomorrow into the arms of today, showing society where it is going, and can take a turn with yesterday to show where it has come from (Griffiths 2009, 6).

Sitting in the small darkened room, the high-energy soundtrack bombards my senses, making my hear beat faster as the frantic rhythm increases its tempo. On screen, the images move quickly from one disaster movie scene to another, as humans, buildings and landscapes are destroyed by hurricanes, floods, meteors, and unidentified objects. As the disasters unfold, I am compelled to watch, searching to recognize one of the films, yet equally horrified by my visceral pleasure in the combination of sound and image. My senses are heightened but satisfaction is deferred by the abrupt ending of the video montage, to be played in a continual, endless loop. I walk away, but come back later to watch it, again, and again, enjoying the assault on my senses, simultaneously aware of its seduction. I feel uncomfortable. Watching Tracey Moffatt's *Doomed* (2007) at the *Earth* exhibition is an embodied experience.

In *Doomed*, sequences from Hollywood disaster films, whose narratives and images are familiar to many of us, are spliced together so that one scene moves into another, mixing old films with new films to create a history of the disaster genre, without a linear timeline. The narrative sequence does not allow resolution, but locks us into a cycle of disaster that seems to build to a crescendo through the accompanying soundtrack, but is deferred by the visual narrative's lack of closure. This makes it both comical and absurd – as the expectation of resolution amplified by the sound is undercut by the continuous present of the visual sequence. While Hollywood's ongoing obsession with disaster is comically revealed, at the same time, these images of disaster take on new meanings in the context of climate change. If we have been able to visualise disaster all along, then why have these imaginings not translated into efforts to avert climate crisis? Are these images, or metaphors, now translating into reality?[1] Do these images work to continuously defer acknowledgement of the real disasters that await us? Are we only equipped to watch our own impending doom (Cottle 2008)? How do we intervene in a narrative for which we do not know the ending? Moffatt's work raises these questions in a chilling way.

1 My thanks to David Harradine for pointing this out to me.

Visual aesthetics/visual epistemology

The delicate clay flowers, intricate in detail, are contained in a tall glass-fronted wooden display cabinet, out of touch. Yet, the doors to the cabinet are partially open, with some flowers resting on the ledge outside, as others lay strewn across the gallery floor, or rest on the mantelpiece of a fireplace on another wall. Many of the flowers are broken; pieces of clay scattered around. I step over them, and around them, the urge to crush and touch palpable. Are we allowed to? Are we being invited to? The gallery room is grand, with a stately home feel. Yet, the fireplace and wall cabinet also add a sense of domesticity familiar to most western homes.

Claire Twomey's *Specimen* (2009) at the *Earth* exhibition does not immediately make me think of climate change. I respond to its beauty, and then the questioning of this beauty through the broken flowers. The wall cabinet makes me think of displays of material wealth in stately houses, via possessions such as vases, crockery and ornaments. I also think of display cabinets in medical museums: surgical tools, stethoscopes, microscopes and body parts; or museum cabinets of stuffed animals; still, inert, dead. I think of the scientific impulse to obtain, classify, organise, know, own. I think about the human desire to display things, to understand and gain knowledge through sight, the visible, the objective, the object. Curiosity/curiosa. I think about an item/object/artefact being used to represent the whole, a part for a whole. Reduction. I am thinking about knowledge and epistemology, what it means to know. The futility of knowing through the acquisition of objects. I am thinking about what it is to be human, what we value, how we act and respond. I am starting to think about climate change.

I come across Twomey's *Specimen* again a few months later at the *Unfold* exhibition. Here it is displayed differently. Occupying the centre of the room, the unfired clay flowers are contained in an open wooden box. The box sits on a small table covered by black cloth which spreads across the floor. Some of the flowers look like they have tumbled out of the box, coming to lie on top of the cloth. Many of them are broken. The feeling is much more intimate than that conveyed at the *Earth* exhibition. I sense that I am looking into a private collection of specimens not yet intended for public visual display. I feel an odd sense of despair and am not sure why. On the inside of the box lid is a genealogical timeline moving from left to right. Starting in 1952 it marks the birth of Pat Martin in Stoke-on-Trent, UK, tracing his training and developing career as a china maker, up to 2009 and his work at Aynsley China, UK. The timeline continues on to an unspecified future. The accompanying exhibition card explains that the traditional skills of china flower making are being lost. This explanation forecloses the work's meaning for me. While it gives some detail to the sense of despair I felt, now made meaningful through the loss of human skills, at the same time this makes the work too specific, losing the wider questioning of knowledge and ownership conveyed by the previous exhibition. I resist this interpretation. I look at the artwork again.

I start to think about the personal histories or stories of those who make artefacts. The domesticity invoked in the work at the *Earth* exhibition is made more apparent here without the grand surroundings. The small china flowers could be ornaments intended for a domestic mass market, to be displayed on people's mantelpieces, in their small display cabinets, on dressing tables, on window sills. This is about mass consumer culture and the loss of specialist production skills, as the fetishism of the commodity takes precedent over concern for who has produced the commodity and under what conditions. The impulse to own, acquire and visually display has translated into a domestic consumer market, with the attendant loss of knowledge of origins. Consumption over production. I am starting to think about climate change again.

Twomey's work illustrates the productiveness of art that is not explicit in its message, but rather forces us to think, reflect and actively reinterpret. My reading of her work may not be her intended one, but it has forced me to search for my own meaning. This is directed by the context of display – two different pieces of work, with the same title, both displayed in exhibitions concerned with cultural responses to climate change. Twomey's *Specimen* was the result of an expedition to the Arctic with Cape Farewell. Yet, I do not read the Arctic in her work as I have in other work from Cape Farewell's projects, such as David Buckland's projections of text (such as *Burning Ice*, 2005) directly onto the Arctic glaciers. Twomey's work is more subtle, less referential. It invites different meanings. The fragility of the Arctic ice may have translated into the fragility of the clay flowers, but her work asks us to more specifically question ways of knowing. In using the beauty of objects she asks us to question those epistemologies of knowledge through which objects and knowledge are acquired, owned and revered. The ultimate futility and limits of these ways of knowing are exposed in the context of climate change.

Spaces of feeling, spaces of display

Earth exhibition

Part of how we experience and respond to artworks is shaped by where we encounter them, through their spaces of display and consumption. The *Earth* exhibition was housed in a traditional art institutional setting, in Burlington House at the Royal Academy of Arts, originally founded in 1768 (Royal Academy of Arts 2010). Even before entering the building, our experiences are influenced by our journey there. On arriving at the Royal Academy, I realise I have come to the wrong entrance, and to get to the other side of the building I have to walk through Burlington Arcade – 'Britain's very first shopping arcade, opened in 1819' – which runs the length of the Academy buildings (Burlington Arcade 2010). A nineteenth century paean to a burgeoning consumer culture, today the window displays are no less conspicuous or luxurious in their presentation of commodities. Through my body, I feel the sudden change from the busy, noisy, rain drenched street outside, as I enter the quiet, dry, brightly lit arcade. I tread the carpeted walkway, visually assaulted by the lights and opulence of the commodity displays. Before I have reached the end of the arcade I am already thinking of how ironic it is that in order to get to an exhibition that seeks to engage us with climate change, I have to experience a space that is the epitome of an energy profligate consumerist lifestyle. At the same time, the classed spaces of the institution I am about to enter are strangely mirrored by the classed space of the arcade, with its tasteful high-market goods. The arcade, I think, should simply be transported into the space of the exhibition as an artwork. It reminds me that is through the unexpected, the incongruous juxtaposition and navigation of different spaces in our everyday lives, that we are forced to think, reflect and question.

Unfold exhibition

I am irritated. I've come to Kings Place in London to visit Cape Farewell's *Unfold* exhibition and I can't find it. The large glass building is an arts and music space, housing galleries, theatres and offices. The lack of signage, and its multiple levels, makes the exhibition space difficult to find. I'm told to go down the escalators, which leads me to the basement bar (empty) and to a few perspex covered photographs on the walls. The series of photographs document the artists who have undertaken a Cape Farewell expedition to the Arctic, and include a short statement of their experience. Is this it? I look up and see further images on the wall in the mezzanine floor above. A sign says 'Gallery'. I go up the stairs and come across a pile of wooden crates stacked on top of each other. Over in the corner is a TV screen with flickering images of glaciers and people. Is this it? The boxes are an artwork by Sam Collins, whose title, *Sometimes the Journey is Better than the Destination* (2010), makes me stop and ponder. Is this the message for me? The getting here, the irritation felt, is this better than finding the artworks? Is this the point of art, to make me question my expectations and assumptions. Maybe. But I still want to see the work.

Collins' work offers a partial antidote to my annoyance. The crates represent the transportation of the artworks, revealing the travel and energy involved in the production, dissemination and consumption of art. A computer screen on top of one of the crates tracks the imaginary movement of the artworks via images of Google Earth. It reveals one aspect of the commodity relations involved in the production of artworks – the physical transportation of goods. The work, itself a commodity, asks us to recognize 'that the arts share an industrial base with other sectors in the global economy' (Tickell 2009, 4). At the same time, the work denies the viewer a glimpse of the final product; here the crates are the artwork. I turn around and see another 'Gallery' sign above the door of a room. I walk over. This is where the main exhibition is. There is one other person in the space. I walk in.

Art's meanings cannot be contained, because they are dependent upon the context of display, and shaped by the viewer's own prior assumptions, perceptions or experiences. This is what makes art both transformative and limiting. Bunting argues that, 'The visual arts reaches a particular audience, and many of them are already engaged in this [climate change] issue; it's not clear how or whether it can reach new audiences' (Bunting 2010). Who is art's intended audience? If art remains the preserve of institutional spaces, who is included/excluded from these spaces? Yet, Bunting's reflections also presume that the knowledge and practices of those people already engaged with the issue of climate change cannot be further transformed. I have been 'engaged' with the issue of climate change since the mid 1990s, yet I am continuously questioning my own perceptions, values and practices. This doesn't stop. Even the irritated and angry moments are productive. Different spaces of display are crucial. I would like to encounter an artwork unexpectedly in my everyday life. I would like to be forced to think and reflect at a time when I am least expecting it. Might this be a way of reaching a wider audience?

Function of climate art?

> [O]ne salient image, sculpture or event can speak louder than volumes of scientific data and engage the public's imagination in an immediate way (Cape Farewell 2010b).

While visual artists play an important cultural role in creating new ways of engaging with climate change, I would disagree with Cape Farewell's claim that one image has the capacity to engage the public in an immediate way. Indeed, while the iconic image of the polar bear certainly captured people's imagination, this image has also made it difficult for climate change to be understood as a human concern in the present, instead of a distant problem for the future. It would therefore be misguided, as well as an impossible task, to attempt to reduce the complexities of climate change, and audiences' engagement with climate art, to one image. A more productive approach would be to acknowledge the multiplicity of ways that art can facilitate people's understanding of climate change. Alison Tickell, Director of Julie's Bicycle, a not for profit company which works to tackle climate change in the creative industries, says, 'Climate change cannot be resolved through the celebration of a single individual's efforts, nor the gallant actions of a few, but through the collective efforts of many' (Julie's Bicycle 2009, 4). This means taking into account the range of audiences that comprise 'the public': an approach that is particularly important given the institutional structures through which many visual artworks are displayed, and the need to interrogate the power relations which permeate these viewing spaces.

Artists face a difficult task: to what extent should they seek to mobilise behavioural change through the communication of a message; or provide us with imaginative spaces through which to reflect upon our role of humans in a climatically changing world (Bunting 2010)? As Bunting argues, 'The visual arts offer a myriad of powerful ways to think and feel more deeply about our age and our humanity, but it is almost impossible to trace the causal links of how that may feed through to political engagement or behaviour change' (Bunting 2010). As such, an intentional message and explicit outcome should not be an expectation of arts engagement with climate change (ibid). Indeed, my own responses to art and climate change have been strongest when existing art works are reinterpreted through the lens of climate change (e.g. Gormley and Hatoum), or when art that may have been inspired by climate change is not instrumental or obvious in its message, instead seeking to make us question what it means to be human, and revealing the limitations of the knowledge systems through which we make sense of the world (e.g. Twomey) in the hope of generating new social realities.

If climate change is not merely to provide 'art with a new narrative' (RSA Arts and Ecology 2010), then collaborative projects between artists and different disciplines or groups of people may facilitate more meaningful and diverse cultural responses to climate change, moving images of climate change away from the obvious to help facilitate more meaningful, present and human centred responses to climate change. Collaborations can be challenging, but in their very process are the means through which new knowledge and understanding is created. Art needs to help open up new spaces of viewing and new modes of understanding.

Conclusion – new presents, new futures

I want to finish this chapter by looking forward. In exploring notions of belonging, place, space, time, vision, display and context through art, I have called attention to the possibilities that art offers for revealing the limitations of existing knowledge systems and for creating new ways of understanding and responding to climate change, more attentive to our embodied, imaginative and emotional experiences. These concerns are also being explored through a creative dialogue I am currently engaged in with the artist, David Harradine. Through a *Leverhulme Trust* funded Artist in Residence Project, which brings an artist into creative collaboration with staff from a non arts academic department, we are exploring how visual arts can better engage people on climate change, through a consideration of the concepts of time, space/place/landscape and embodiment (Here Today 2010). In engaging with these concepts we are both calling attention to the problems of current visualisations of climate change in popular media culture, which have tended to focus upon climate impacts, and seeking to offer new ways of perceiving climate change through a creative and critical engagement with these concepts.

As a media studies academic, I bring with me an understanding of the difficulties climate change poses for visual communication, and a desire to move away from images of climate impacts to try to find a new visual language for climate change. As a performance maker and visual artist, Harradine brings an expertise in using visual and spatial language to communicate and explore through practice, concepts of light, landscape, ecology, weather and place (Fevered Sleep 2010). Sharing both a dislike of polar bear images, and a concern for a more embodied and emotional engagement with climate change, we are not seeking to replace one image with another, but rather to create a conceptual framework through which to explore and offer new ways of visualising climate change. This is not an easy task, particularly given our different disciplinary perspectives. Yet, through collaborative dialogue we have attempted to define the limits and possibilities of the artwork we will eventually produce, through an ongoing manifesto:

It must not be literal
It must make me feel something
It must not be confined to an art space
It must not depict a utopia
It must not depict a dystopia
It must have an element of sound
It must complicate our experience of time
It must include the human
It must not contain images in which polar bears appear to be cute
It must be playful
It must be deadly serious
It must not be singular
It must not use un(re)contextualised images of climate impacts
It must consider its audience
It must appear in more than one form in more than one place
It must be mobile

To offer the possibilities for more meaningful engagements with climate change, we need firstly to understand what the challenges for communication and understanding are, in the hope of finding a new (visual) language for climate change. Our project, *Here Today: Moving Images of Climate Change*, is a small attempt to do so (Here Today 2010). In contributing to a developing field of visual arts responses to climate change, it seeks to add a critically informed perspective to a multitude of growing creative voices, to help make climate change resonate as a future present, rather than a future possible.

Epilogue
Positive Action in a Changing Climate

Recent print advertisements from the UK Government's Act on CO2 climate campaign (2009) attempt to give cultural meaning to climate change by rewriting children's nursery rhymes to subvert their original message. 'Twinkle, Twinkle little star, how I wonder what you are', is revised to continue: 'up above the world so high, looking down at dangerously high levels of CO2 in the atmosphere' (UK Government 2009). In another example, 'Jack and Jill went up the hill to fetch a pail of water', continues: 'There was none, as extreme drought due to climate change had caused a drought' (ibid). The advertisements play with concepts of time and cultural memory. The focus upon children is a way of bringing the impacts of future climate change into the present, while the use of traditional line drawings to illustrate the text evokes the nostalgia of childhood memories, intensifying the feelings of (actual and potential) loss for the adult audience. The advertisements look forward to the future and hark back to the past in order to make climate change meaningful in the present.

Communicating a negative message, the advertisements are quite crude in their approach, activating fear and guilt to promote behavioural change. While these forms of communication serve to disengage people (Moser and Dilling 2004), of particular significance was the public opposition to the content and tone of a television advertisement in the same campaign. Featuring a man reading a bedtime story to a young girl, the story is set in a land where 'the weather was very very strange', with heat-waves, storms and floods (UK Government 2009). The reason for this weather is explained in the story by scientists who 'said it was being caused by too much CO2, which went up into the sky when the grown ups used energy…And it was the children of the land who'd have to live with the horrible consequences' (ibid). The advertisement ends with the young girl asking if there is a happy ending, with the voice over stating, 'It's up to us how the story ends' (ibid). Over 1,000 complaints were made to the UK Advertising Standards Authority and to the communications regulator Ofcom, on the basis that the advertisement should not have attributed climate change to human activity as the science is inconclusive, and that the tone was upsetting (Sweney 2010). The UK Government defended the advertisements by claiming that it 'was the first time the government had thrown its weight behind the view that scientific evidence now clearly shows that climate change is man made' (ibid). Although the number of complaints was relatively small, they provide a snapshot of the climate of public opinion in the UK, and the reticence of governments to address climate change, who are only now throwing 'their weight' behind science (ibid). They illustrate the ongoing difficulties encountered in communicating the science of climate change and in making it culturally meaningful and personally relevant in order to promote action.

Mediating Climate Change has explored the problems climate change has posed for knowledge, communication and action. Temporally complex and not always visible, climate change presents difficulties for the scientific knowledge systems and the representational practices called upon to identify, communicate and make it meaningful. Central to these difficulties are the interrelated conceptions of nature, vision and time, as they are constructed through and by science, which have figured the environment as a visible, objective and commodified nature (Beck 1992; Adam 1988a, 1988b; Plumwood 2002). Such conceptions have been historically reinforced through mainstream environmentalism and mediated communication, which have conversely served to present humans as separate and disconnected from the environment, through the prioritisation of a visual aesthetics which subscribes to a visible present/past. As such, this book has called for a reconception of the environment, and of climate change, as the imbricated and mutually interdependent relationship between humans and ecosystems.

Through an analysis of contemporary mediated communication on climate change, including an emerging climate movement, print news media, civil society groups and visual artists, this book has also explored how climate change is being made culturally meaningful through a range of discursive and social practices. It found that climate change is variously framed through appeals to discourses of justice, faith, scientific certainty, ethics, emotion and morality to engage people on climate change and promote positive action. At the same time, discourses of scientific uncertainty, neoliberalism, (ir)rationality, individualism and anti-environmentalism have been used to promote inaction. There is not one way to communicate climate change more effectively, nor one form of action to be undertaken. However, communication on climate change needs to be compelling and convey the urgency of this issue (Moser and Dilling 2004; Risbey 2007). In doing so, it needs to explain the temporal, invisible and incremental nature of climate change, as well as the uncertainties of climate science in relation to its magnitude, rapidity of change and geographical specificity, as a basis for action in the present. Communication needs to engage people in ways that are meaningful to them, and to encourage the efficacy of different modes of self and collective action. In letting go of nature, as it has come to be constructed through western knowledge systems, we can hope to connect and engage with climate change in a way that acknowledges its embeddedness in our everyday lives, encouraging a change in our perceptions and our actions.

References

Adam, B. (1994), 'Running out of time: Global crisis and human engagement', in M. Redclift and T. Benton (eds).

Adam, B. (1998a), *Timescapes of Modernity: The Environment and Invisible Hazards* (London: Routledge).

Adam, B. (1998b), 'Values in the cultural timescapes of science', in Scott Lash et al. (eds).

Adam, D. and Jowit, J. (2008), 'People power vital to climate deal – minister', *guardian.co.uk* [website], (8 December 2008), <http://www.guardian.co.uk/environment/2008/dec/08/ed-miliband-climate-politics-environment>, accessed 7 January 2009.

Adams, C.J. (1990), *The Sexual Politics of Meat: A Feminist-Vegetarian Critical Theory* (London: Polity Press).

Allan, S. et al. (eds) (2000), *Environmental Risks and the Media* (London: Routledge).

Allen, M.W. and Baines. S. (2002), 'Manipulating the symbolic meaning of meat to encourage greater acceptance of fruits and vegetables and less proclivity for red and white meat', *Appetite* 38, 118-130.

Anderson, A. (1997), *Media, Culture and the Environment* (London: Routledge).

Anderson, J. (2004), 'Spatial politics in practice: The style and substance of environmental direct action', *Antipode* 36:1, 106-125.

Andreadis, E. and Smith, J. (2007), 'Beyond the oozone layer', *British Journalism Review* 18:1, <http://www.bjr.org.uk/data/2007/no1_andreadis_smith.htm>, accessed 3 July 2007.

Animal Aid (2010), 'Meat-free goes mainstream', *Outrage* 159 (Summer), 6-7.

Anon (2010), 'An open letter to the climate camp neighbourhoods', in Camp For Climate Action (2010b).

Antilla, L. (2005), 'Climate of scepticism: US newspaper coverage of the science of climate change', *Global Environmental Change* 15:4, 338-352.

Assadourian, E. (2010), 'The rise and fall of consumer cultures', in WorldWatch Institute.

Barnett, C. et al. (2004), 'Philosophy and ethical consumption', in R. Harrison et al. (eds).

Barthes, R. (1977), *Image, Music, Text* (London: Fontana Press).

Barthes, R. ([1980] 2000), *Camera Lucida: Reflections on Photography* (London: Vintage).

Bauman, Z. (2001), *The Individualized Society* (Cambridge: Polity Press).

Bauman, Z. (2007), *Consuming Life* (Cambridge: Polity Press).

BBC/Open University (1991), *The Heat is On* [Television Documentary].
BBC (1997), *The Science of Climate* [Television Documentary].
BBC (2000), *Underwater Britain,* [Television Documentary].
BBC (2006a), *Are We Changing Planet Earth?* [Television Documentary].
BBC (2006b), *Can We Save Planet Earth?* [Television Documentary].
BBC (2009a), 'What is the BBC?' [website], <http://www.bbc.co.uk/info/purpose/what.shtml>, accessed 27 July 2009.
BBC (2009b), 'Missions and values' [website], <http://www.bbc.co.uk/info/purpose/index.shtml>, accessed 27 July 2009.
BBC (2009c), 'Public purposes' [website], <http://www.bbc.co.uk/info/purpose/public_purposes/index.shtml>, accessed 27 July 2009.
Beck, U. (1992), *The Risk Society: Towards a New Modernity* (London: Sage).
Bell, A. (1994), 'Climate of opinion: Public, media discourse on the global environment', *Discourse and Society* 5:1, 33-64.
Bickerstaff, K. et al. (2006), 'Reframing nuclear power in the UK energy debate: nuclear power, climate change mitigation and radioactive waste', *Public Understanding of Science* 15, 1-25.
Blain, L. (2005), 'Ain't I an Environmentalist?: An environmental-justice advocate insists he's not dead yet', *Grist*, (31 May 2005), <http://www.grist.org/article/blain-death>, accessed 15 January 2010.
Boehnert, J. (2008), 'Design in the front line', *Creative Review*, (31 October 2008), <http://www.creativereview.co.uk/cr-blog/2008/october/design-in-the-front-line>, accessed 4 May 2009.
Bourdieu, P. (1984), *Distinction: A Social Critique of the Judgement of Taste* (London: Routledge).
Boyce, T. and Lewis, J. (eds) (2009), *Climate Change and the Media* (New York: Peter Lang).
Boykoff, M.T. (2007), 'Flogging a dead norm? Newspaper coverage of anthropogenic climate change in the United States and United Kingdom from 2003 to 2006', *Area* 39:4, 470-481.
Boykoff, M.T. (2008), 'The cultural politics of climate change discourse in the UK tabloids', *Political Geography* 27, 549-569.
Boykoff, M.T. (2009), 'We speak for the trees: Media reporting on the environment', *Annual Review of Environment and Resources* 34, 431-458.
Boykoff, M.T. and Boykoff, J.M. (2004), 'Balance as bias: Global warming and the US prestige press', *Global Environmental Change* 14, 125-136.
Boykoff, M.T. and Roberts. J.T. (2007), 'Media coverage of climate change: Current trends, strengths, weaknesses', *United Nations Development Programme – Human Development Report 2007 – Background Paper.*
Boykoff, M.T. and Mansfield, M. (2008), '"Ye Olde Hot Aire": Reporting on human contributions to climate change in the UK Environmental Press', *Environmental Research Letters* 3, 1-8.

Boykoff, M. and Goodman, M.K. (2009), 'Conspicuous redemption? Reflections on the promises and perils of the "Celebritization" of climate change', *Geoforum* 40, 395-406.

Braasch, G. (2005), 'World view of global warming: The photographic documentation of climate change' [website], <http://www.worldviewofglobalwarming.org/index.html> (home page), accessed 18 December 2007.

Brockington, D. (2008), 'Powerful environmentalisms: conservation, celebrity and capitalism', *Media, Culture and Society* 34:4, 551-568.

Brown, A. (2005), 'Rainbow warrior: Dramatizing the "death" of environmentalism doesn't help urban people of color, or anyone else', *Grist*, (15 March 2005), <http://www.grist.org/article/brown8>, accessed 15 January 2010.

Bryant, R.L. and Goodman, M.K. (2004), 'Consuming narratives: The political ecology of "alternative" consumption', *Transactions of the Institute of British Geographers* 29, 344-366.

Bucchi, M. (1998), *Science and the Media: Alternative Routes in Scientific Communication* (London: Routledge).

Budescu, D. et al. (2009), 'Improving communication of uncertainty in the reports of the Intergovernmental Panel on Climate Change', *Psychological Science* 20:3, 299-308.

Bunting, M. (2010), 'An essay: Does the art stop when the persuasion starts?', *RSA Arts and Ecology* [website],<http://www.artsandecology.org.uk/magazine/features/madeleine-bunting>, accessed 16 May 2010.

Burlington Arcade (2010), 'About the arcade' [website], <http://www.burlington-arcade.co.uk/main.php?id=1>, accessed 16 June 2010.

CAFOD (2009a), 'General questions about Cafod' [website], <http://www.cafod.org.uk/about-us/faqs/general-questions>, accessed 12 December 2009.

CAFOD (2009b), 'Vision, Mission and Values' [website], <http://www.cafod.org.uk/about-us/vision-mission-values>, accessed 12 December 2009.

CAFOD (2009c), 'World Catholics demand climate justice' [website], <http://www.cafod.org.uk/news/climate-justice-2008-12-08>, accessed 1 January 2010.

CAFOD (2010a), 'Tricky questions and answers on the Climate Justice campaign' [website], <http://www.cafod.org.uk/policy-campaigns/climate-kit>, accessed 2 February 2010.

CAFOD (2010b), 'Handy hints to grow the Climate Justice campaign' [website], <http://www.cafod.org.uk/policy-campaigns/climate-kit>, accessed 2 February 2010.

Cammaerts, B. (2007), 'Media and communication strategies of glocalised activists: Beyond media-centric thinking', in B. Cammaerts and N. Carpentier (eds).

Cammaerts, B. and Carpentier, N. (eds) (2007), *Reclaiming the Media: Communication Rights and Democratic Media Roles* (Bristol: Intellect Books).

Camp for Climate Action (2008a), 'Aims of the Camp for Climate Action' [website], <http://www.climatecamp.org.uk/aims.php>, accessed 26 April 2008.

Camp for Climate Action (2008b), 'The Camp for Climate Action 2007' [website], <http://www.climatecamp.org.uk/about.php>, accessed 26 April 2008.

Camp for Climate Action (2009a), 'G20 in the city '09' [website], <http://www.climatecamp.org.uk/actions/g20-2009>, accessed 16 November 2009.

Camp for Climate Action (2009b), 'Camp for Climate Action '09' [website], <http://www.climatecamp.org.uk/actions/london-2009>, accessed 16 November 2009.

Camp for Climate Action (2010a), 'About us' [website], <http://www.climatecamp.org.uk/about>, accessed 7 February 2010.

Camp for Climate Action (2010b), *Criticism without Critique: A Climate Camp Reader* (London: Shift Magazine/dysophia production) <http://dysophia.files.wordpress.com/2010/01/cca_reader.pdf>, accessed 16 June 2010.

Cape Farewell (2010a), 'About us' [website], <http://www.capefarewell.com/about.html>, accessed 17 June 2010.

Cape Farewell (2010b), 'The art programme' [website], <http://www.capefarewell.com/art.html>, accessed 17 August 2010.

Caritas Internationalis (2009), 'About Caritas' [website], <http://www.caritas.org/about/Caritas_Internationalis.html>, accessed 5 January 2010.

Carvalho, A. and Burgess, J. (2005), 'Cultural circuits of climate change in U.K. Broadsheet Newspapers, 1985-2003', *Risk Analysis* 25:6, 1457-1469.

Carvalho, A. (2007), 'Ideological cultures and media discourses on scientific knowledge: re-reading news on climate change', *Public Understanding of Science* 16, 223-243.

Channel 4. (2009), Channel 4 News, 19.15, 28 March 2009.

Chapman, J. (2005), *Emotionally Durable Design: Objects Experiences and Empathy* (London: Earthscan).

Chouliaraki, L. (2008), 'The media as moral education: mediation and action', *Media Culture and Society* 30:6, 831-852.

CIDSE Bishop's Statement (2008), 'A call for climate justice', <http://www.cafod.org.uk/news/climate-justice-2008-12-08/panels/related-resources/cidse-bishops-statement>, accessed 14 January 2010.

Climate Action Network (CAN) (2009), [website], <http://www.climatenetwork.org/>, accessed 15 December 2009.

Climate Justice Now! (2009), [website], <http://www.climate-justice-now.org/>, accessed 15 December 2009.

Clarke, N. (2008), 'From ethical consumerism to political consumption', *Geography Compass* 2:6, 1870-1884.

Compassion in World Farming and Friends of the Earth (2009), *Eating the Planet? How We can Feed the World Without Trashing it* (London: CIWF and FoE).

Connor, S. (2006), 'Scientist publishes 'escape route' from global warming', *The Independent*, (31 July 2006), <http://www.independent.co.uk/environment/scientist-publishes-escape-route-from-global-warming-409981.html>, accessed 18 February 2009.

Corbett, J.T. (2006), *Communicating Nature: How We Create and Understand Environmental Messages* (Washington, DC: Island Press).

Cosgrove, D. (1994), 'Contested Global Visions: One World, Whole-Earth, and the Apollo Space Photographs', *Annals of the Association of American Geographers* 84:2, 270-294.

Cottle, S. (2000), 'TV News, lay voices and the visualization of environmental risks' in S. Allan et al. (eds).

Cottle, S. (2008), 'Reporting demonstrations: The changing media politics of dissent', *Media, Culture and Society* 30:6, 853-872.

Cottle, S. (2009), *Global Crisis Reporting: Journalism in the Global Age* (Maidenhead: Open University Press).

Cottle, S and Nolan, D. (2009), 'How the media's codes and rules influence the way NGOs work', *Nieman Journalism Lab*, (16 November 2009), <http://www.niemanlab.org/2009/11/simon-cottle-and-david-nolan-how-the-medias-codes-and-rules-influence-the-ways-ngos-work/>, accessed 20 December 2009.

Cox, R. (2006), *Environmental Communication and the Public Sphere* (London: Sage).

Crompton, T. and Kasser, T. (2009), *Meeting Environmental Challenges: The Role of Human Identity* (Goldalming: WWF-UK).

Della Porta, D. and Tarrow, S. (eds) (2005), *Transnational Protest and Global Activism* (Oxford: Rowman and Littlefield Publishers).

Della Porta, D. and Diani, M. (2006), *Social Movements: An Introduction* (Oxford: Blackwell Publishing).

DeLuca, K.M. (1999), *Image Politics: The New Rhetoric of Environmental Activism* (New Jersey: Lawrence Erlbaum Associates).

Diani, M. and Donati, P.R. (1999), 'Organisational change in Western European wnvironmental groups: A framework for analysis', in C. Rootes (ed).

Dobrin, S. and Morey, S. (eds), *Ecosee: Image, Rhetoric, and Nature* (New York, NY: State University of New York Press).

Doherty, B. (2006), 'Friends of the Earth International: Negotiating a transnational identity', *Environmental Politics* 15:5, 860-880.

Doyle, J. (2001), *Imag(in)ing Anatomy: Dissecting Difference Through the Art of Surgery*, Unpublished Doctoral Thesis, University of Sussex, UK.

Doyle, J. (2007), 'Picturing the clima(c)tic: Greenpeace and the representational politics of climate change', *Science as Culture* 16:2, 129-150.

Doyle, J. (2009), 'Seeing the climate? The problematic status of visual evidence in climate change campaigning', in S. Dobrin and S. Morey (eds).

Doyle, J. (2009). 'Climate action and environmental activism: The role of environmental NGOs and grassroots movements in the global politics of climate change' in T. Boyce and J. Lewis (eds).

Doyle, J. (2011), 'Acclimatising nuclear? Climate change, nuclear power and the reframing of risk in the UK news media', *International Communication Gazette* 73:1-2, 107-125.

Doyle, T. and McEachern, D. (2008), *Environment and Politics* (London: Routledge).

Elliot, L. (1998), *The Global Politics of the Environment* (Basingstoke: Macmillan).

Elliott, J. (2006), *An Introduction to Sustainable Development* (London: Routledge).

European Commission (2005), *EU action against climate change, EU emissions trading: an open scheme promoting global innovation* (Brussels: European Commission), <http://ec.europa.eu/environment/climat/pdf/emission_trading2_en.pdf>, accessed 7 June 2008.

European Commission (2009), 'You control climate change', [website] <http://ec.europa.eu/environment/climat/campaign/>, accessed 18 December 2009.

Farrow, C. (2000), 'Communicating about climate change: an NGO view', in J. Smith (ed.).

Fevered Sleep (2010), [website], <http://www.feveredsleep.co.uk/>, accessed 18 August 2010.

Fenton, N. (2009), 'Has the Internet changed how NGOs work with established media? Not enough', *Nieman Journalism Lab*, (23 November 2009), <http://www.niemanlab.org/2009/11/natalie-fenton-has-the-internet-changed-how-ngos-work-with-established-media-not-enough/>, accessed 20 December 2009.

FoE International (2010a), 'Who we are' [website], <http://www.foei.org/en/who-we-are>, accessed 17 May 2010.

FoE International (2010b), 'What we do' [website], <http://www.foei.org/en/what-we-do>, accessed 4 January 2010.

FoE International (2010c), 'Funding and membership support' [website], <http://www.foei.org/en/resources/publications/annual-report/2008/financial-report/funding-and-membership-support>, accessed 4 January 2010.

FoE International (2010d), 'Affected people and impacted communities' [website], <http://www.foei.org/en/what-we-do/affected-peoples>, accessed 4 January 2010.

FoE International (2010e), 'Climate justice and energy' [website], <http://www.foei.org/en/what-we-do/climate-and-energy>, accessed 4 January 2010.

FoE International (2010f), 'Our climate demands' [website], <http://www.foei.org/en/what-we-do/un-climate-talks/global/2009/our-demands-in-copenhagen/?searchterm=our%20climate%20demands>, accessed 4 January 2010.

FoE International (2010g), 'Grassroots highlights – affected people and impacted communities' [website], <http://www.foei.org/en/what-we-do/affected-peoples/grassroots-highlights>, accessed 4 January 2010.

FoE international (2010h), 'Declaration from the movement of victims affected by climate change' [website], <http://www.foei.org/en/what-we-do/affected-peoples/grassroots-highlights/2009/declaration-from-the-movement-of-victims-affected-by-climate-change>, accessed 4 January 2010.

FoE UK (2008), *What's Feeding Our Food? The Environmental and Social Impacts of the Livestock Sector* (London: Friends of the Earth).

FoE UK (2009), 'Fix the Food Chain', [website], <http://www.foe.co.uk/what_we_do/fix_food_chain_20576.html>, accessed 19 June 2010.

FoE UK (2010a), 'Climate change – too risky to ignore' [website], <http://www.foe.co.uk/campaigns/climate/news/climate_change_23261.html>, accessed 31 March 2010.

FoE UK (2010b), 'Current campaigns: Get serious about CO2' [website], <http://www.foe.co.uk/campaigns/climate/get_serious/>, accessed 31 March 2010.

FoE UK (2010c), 'Get serious about CO2' [website], <http://www.foe.co.uk/serious_about_co2/index.html>, accessed 31 March 2010.

Food and Agriculture Organisation of the United Nations, FAO (2006), *Livestock's Long Shadow: Environmental Issues and Options* (Rome: Italy).

Foucault, M. ([1970] 2002). *The Order of Things: An Archaeology of the Human Sciences* (London: Routledge).

Fox, I. (2009), 'London on parade to show that ethical clothes can cut it on the catwalk', *The Guardian* (21 February 2009), 13.

Franklin et al. (2000), *Global Nature, Global Culture* (London: Sage).

Galtung, J. and Ruge, M. (1965), 'The structure of foreign news: The presentation of the Congo, Cuba and Cyprus crises in four Norwegian newspapers', *Journal of International Peace Research* 1, 64-91.

Gelbspan, R. (1997), *The Heat is On: The High Stakes Battle Over Earth's Threatened Climate* (Reading, MA: Addison Wesley).

Gelobter, M. et al. (2005), 'Standing on whose shoulders? Why race and class matter to the environmental movement', *Grist* (27 May 2005), <http://www.grist.org/article/gelobter-soul>, accessed 15 June 2009.

Gilbert, J. (2008), 'Against the commodification of everything: Anti-consumerist cultural studies in the age of ecological crisis', *Cultural Studies* 22:5, 551-566.

Green Alliance (2008), *The New Politics of Climate Change: Why We Are Failing and How We Will Succeed* (London: Green Alliance).

Greenpeace UK (1990), *Climate Change: A Reader's Guide to the IPCC Reports* (London: Canonbury Villas).

Greenpeace UK (1997), *Putting the Lid on Fossil Fuels* (London: Greenpeace UK).

Greenpeace UK (1998), *Nature's Bottom Line: Protection and the Carbon Logic* (London: Greenpeace UK).

Greenpeace UK (2001), 'A decade of dirty tricks: ExxonMobil's attempts to stop the world tackling climate change' (London: Greenpeace UK).

Greenpeace UK (2002), *Speak Out!* (London: Greenpeace UK).

Greenpeace UK (2004), *Sea, Wind, Europe* (London: Greenpeace UK).

Greenpeace International (1993), *Emerging Impacts of Climate Change?* (Amsterdam: Greenpeace International).

Greenpeace International (1999), 'Arctic expedition' [website], <http://archive.greenpeace.org/climate/arctic99/indexb.html>, accessed 16 June 2009.

Greenpeace International (2002), 'Exxon's weapons of mass deception' [website], <http://www.greenpeace.org.uk/files/pdfs/migrated/MultimediaFiles/Live/FullReport/5292.pdf>, accessed 16 June 2009.

Greenpeace International (2007), *How to Save the Climate: Join the Energy [R]evolution*, Greenpeace International [website], <http://www.greenpeace.org.uk/files/pdfs/climate/howtosavetheclimatepers.pdf>, accessed 18 June 2009.

Greenpeace International (2010), 'Ask Nestle to give rainforests a break' [website], <http://www.greenpeace.org/international/campaigns/climate-change/kitkat/>, accessed 4 May 2010.

Griffiths, J. (2009), 'The far-seers of art', in Julie's Bicycle (eds).

g.r.o.a.t (2010), 'Critiquing climate camp', in Camp for Climate Action (2010b).

Guthman, J. (2008), 'Bringing good food to others: investigating the subjects of alternative food practice', *Cultural Geographies* 15:4, 431-447.

Hajer, M.A. (1995), *Politics of Environmental Discourse: Ecological Modernization and the Policy Process* (Oxford: Oxford University Press).

Hale, S. (2008), *The New Politics of Climate Change: Why We are Failing and How We Will Succeed* (London: Green Alliance).

Hansen, A. (1991), 'The media and the social construction of the environment', *Media, Culture and Society* 13, 443-456.

Hansen, A. (ed.) (1993), *The Mass Media and Environmental Issues* (Leicester: The University of Leicester Press).

Hansen, A. (1993), 'Greenpeace and press coverage of environmental issues', in A. Hansen (ed.).

Hansen, A. (2000), 'Claims-making and framing in British newspaper coverage of the 'Brent Spar' controversy', in S. Allan et al. (eds).

Hansen, A. (2010), *Environment, Media and Communication* (London: Routledge).

Hansen, A. and Machin, D. (2008), 'Visually branding the environment: Climate change as a marketing strategy', *Discourse Studies* 10:6, 777-794.

Harcup, T. and O'Neill, D. (2001), 'What is news? Galtung and Ruge revisited', *Journalism Studies* 2:2, 261-280.

Harrison, R. et al. (eds) (2005), *The Ethical Consumer* (London: Sage).

Haynes, J. (1999), 'Power, politics and environmental movements in the third world', in Rootes (ed.) 1999.

Heinz, B. and Lee, R. (1998), 'Getting down to the meat: The symbolic construction of meat consumption', *Communication Studies* 49 (Spring), 86-99.

Here Today (2010), 'Here today: Moving images of climate change' [website], <http://arts.brighton.ac.uk/research/here-today--moving-images-of-climate-change> accessed 18 August 2010.

Hertwich, E.G. and Peters, G.P. (2009), 'Carbon footprint of nations: A global, trade-linked analysis', *Environmental Science and Technology* 43:16, 6414-6420.

High Representative and the European Commission (2008), *Climate Change and International Security*. Paper from the High Representative and the European Commission to the European Council, 14 March 2008.

Hobson, K. (2003), 'Thinking habits into action: The role of knowledge and process in questioning household consumption practices', *Local Environment* 8:1, 95-112.

Hobson, K. (2006), 'Competing discourses of sustainable consumption: Does "rationalization of lifestyles" make sense?', in T. Jackson (ed.).

Hoskins, A. and O'Laughlin, B. (2009), *Television and Terror: Conflicting Times and the Crisis of News Discourse* (Basingstoke: Palgrave Macmillan).

Houghton, J. (2004), *Global Warming: The Complete Briefing* (Cambridge: Cambridge University Press).

Hulme, M. (2008), 'The Conquering of Climate: Discourses of Fear and Their Dissolution', *Geographical Journal* 174:1, 5-16.

Hulme, M. (2009), *Why We Disagree About Climate Change: Understanding Controversy, Inaction and Opportunity* (Cambridge: Cambridge University Press).

Hulme, M. et al. (2009), 'Unstable climates: exploring the statistical and social constructions of "normal" climate', *Geoforum* 40:2, 197-206.

IIED and nef (2004), *Up in Smoke? Threats From, and Responses to, the Impact of Global Warming on Human Development* (London: nef).

Indymedia (2009), 'About us' [website], <http://www.indymedia.org/en/static/about.shtml>, accessed 18 November 2009.

Ingold, T. (1993), 'Globes and Spheres: The Topology of Environmentalism', in *Environmentalism: The View from Anthropology* (ed) Kay Milton (London: Routledge).

IPCC (1990), *Climate Change: The IPCC Scientific Assessment* (Cambridge: Cambrdige University Press).

IPCC (1995), *Climate Change 1995: The Science of Climate Change* (Cambridge: Cambridge University Press).

IPCC (2001a), *Climate Change 2001: The Scientific Basis* [website], <http://www.ipcc.ch/ipccreports/tar/wg1/pdf/WG1_TAR-FRONT.PDF>, accessed 14 June 2009.

IPCC (2001b), *Climate Change 2001: Synthesis Report* (Cambridge: Cambridge University Press).

IPCC (2004), *16 Years of Scientific Assessment in Support of the Climate Convention* [website], <http://www.ipcc.ch/about/anniversarybrochure.pdf>, accessed 18 January 2008.

IPCC (2007a), *Climate Change 2007: The Physical Science Basis* [website], <http://www.ipcc.ch/publications_and_data/publications_ipcc_fourth_assessment_report_wg1_report_the_physical_science_basis.htm>, accessed 1 May 2009.

IPCC (2007b), *Climate Change 2007 Synthesis Report: Contribution of Working Groups I, II and III to the Fourth Assessment Report of the Intergovernmental Panel on Climate Change* [website], <http://www.ipcc.ch/publications_and_data/publications_ipcc_fourth_assessment_report_synthesis_report.htm>, accessed 2 January 2008.

IPPR (2009), *Consumer Power: A Communications Guide for Mainstreaming Lower-Carbon Behaviour* (London: IPPR).

Jackson, T. (2004), 'Negotiating Sustainable Consumption: A Review of the Consumption Debate and its policy implications', *Energy and Environment* 15:6, 1027-1051.

Jackson. T. (ed.) (2006), *The Earthscan Reader in Sustainable Consumption* (London: Earthscan).

Jackson, T. (2008), 'Where is the "wellbeing dividend"? Nature, structure and consumption inequalities', *Local Environment* 13:8, 703-723.

Jamison, A. (1996), 'The Shaping of the Global Environmental Agenda: The Role of Non-Governmental Organisations', in S. Lash et al. (eds).

Jasanoff, S. and Long Martello, L. (eds) (2004), *Earthly Politics: Local and Global in Environmental Governance* (Cambridge, MA: The MIT Press).

Jasanoff, S. (2004), 'Heaven and earth: The politics of environmental images', in S. Jasanoff and M. Long Martello (eds).

Jha, A. and Randerson, S. (2009), 'Poll shows public disquiet about policing at environmental protests', *guardian.co.uk* [website], (25 August 2009), <http://www.guardian.co.uk/environment/2009/aug/25/christian-aid-policing>, accessed 15 November 2009.

Jordan, S. (2010), 'Where next for the Climate Camp?', in Camp for Climate Action (2010b).

Jordanova, L. (1989), *Sexual Visions: Images of Gender in Science and Medicine between the Eighteenth and Twentieth Centuries* (Wisconsin: University of Wisconsin Press).

Jowit, J. (2010), 'Ed Miliband declares war on climate change sceptics', *The Observer* [website], (31 January 2010), <http://www.guardian.co.uk/environment/2010/jan/31/ed-miliband-climate-change-scepticism>, accessed 1 February 2010.

Julie's Bicycle (eds) (2009), *Long Horizons: An Exploration of Art and Climate Change* (London: British Council).

Kegley, C.W and Blanton, S.L. (2009), *World Politics: Trend and Transformation* (Florence, KY: Wadsworth Publishing Company).

Klein, N. (2001), *No Logo: Taking Aim at the Brand Bullies* (London: Flamingo).

Klein, N. (2009), 'Seattle's activism's coming of age will be very disobedient', *The Guardian*, 13 November 2009, 37.

Klimaforum09 (2009), [website], <http://www.klimaforum09.org/?lang=da>, accessed 18 December 2009.

Kolbert, E. (2009). 'Reporting on climate change', in G. Schmidt and J. Wolfe (eds).

Krosnick, J.A. et al (2000), 'The impact of the fall 1997 debate about global warming on American public opinion', *Public Understanding of Science* 9, 239-260.

Lash, S. et. al. (eds) (1996), *Risk, Environment and Modernity: Towards a New Ecology* (London: Sage).

Lash, S, et al. (eds) (1998), *Time and Value* (Oxford: Blackwell Publishers).

Latour, B. (2004), *Politics of Nature: How to Bring the Sciences into Democracy* (Cambridge, MA: Harvard University Press).

Le Gall, D.R. (1996), *Symbols of Catholicism* (Paris: Editions Assouline).

Lean, G. (2004), 'Global warming "will redraw map of world"', *The Independent* [website], (7 November 2004), <http://www.independent.co.uk/environment/global-warming-will-redraw-map-of-world-532305.html>, accessed 17 October 2008.

Leggett, J. (ed) (1990), *Global Warming: Greenpeace Report* (Oxford: Oxford University Press).

Leiserowitz, A. (2006), 'Climate Change Risk Perception and Policy Preferences: The Role of Affect, Imagery and Values', *Climatic Change* 77, 45-72.

Leiserowitz, A. et al. (2008), *Climate Change in the American Mind: Americans' Climate Change Beliefs, Attitudes, Policy Preferences, and Actions* (Washington: George Mason University).

Leiserowitz, A. et al. (2010), *Climate Change in the American Mind: Americans' Global Warming Beliefs and Attitudes in January 2010* (New Haven, CT: Yale University and George Mason University), <http://environment.yale.edu/uploads/AmericansGlobalWarmingBeliefs2010.pdf>, accessed 10 August 2010.

Lester, L. (2007), *Giving Ground: Media and Environmental Conflict in Tasmania* (Hobart: Quintus Publishing).

Lester, L. and Cottle, S. (2009), 'Visualizing climate change: Television news and ecological citizenship', *International Journal of Communication* 3, 920-936.

Lester, L. and Hutchins, B. (2009), 'Power games: environmental protest, news media and the internet', *Media, Culture and Society* 31:4, 579-595.

Linder, S.H. (2006), 'Cashing-in on risk claims: On the for-profit inversion of signifiers for "Global Warming"', *Social Semiotics* 16:1, 103-132.

Littler, J. (2009), *Radical Consumption: Shopping for Change in Contemporary Culture* (Maidenhead: Open University Press).

Longbottom, W. (2008), 'Campaigners arrested in clashes with police over coal power station protest', *The Daily Mail* [website], (4 August 2008), <http://www.dailymail.co.uk/news/article-1041616/Campaigners-arrested-clashes-police-coal-power-station-protest.html#ixzz0wDDKQAmV>, accessed 10 January 2010.

Lorenzoni, I. et al. (2006), 'Cross-national comparisons of image associations with "global warming" and "climate change" among laypeople in the United States of America and Great Britain', *Journal of Risk Research* 9:3, 265-281.

Lupton, D. (1996), *Food, the Body and the Self* (London: Sage).

Macnaghten, P. and Urry, J. (1998), *Contested Natures* (London: Sage).

Manzo, K. (2010), 'Imaging vulnerability: The iconography of climate change' *Area* 42:1, 96-107.

Marshall, G. (2009), 'Leaked email climate smear was a PR disaster for UEA', *The Guardian* [website], (29 November 2009), <http://www.guardian.co.uk/environment/cif-green/2009/nov/23/leaked-email-climate-change>, accessed 1 December 2009.

Mary McCartney (2010), [website] <http://www.marymccartney.net/home.php>, accessed 15 June 2010.

Marx, K. ([1867] 1990), 'The fetishism of commodities and the secret', in *Capital: A Critique of Political Economy, Vol. 1* (London: Penguin).

McAdam, D, et al. (eds), *Comparative Perspectives on Social Movements: Political Opportunities, Mobilizing Structures, and Cultural Framings* (Cambridge: Cambridge University Press).

McAdam, D, et al. (1996), 'Introduction: Opportunities, mobilizing structures, and framing processes – toward a synthetic, comparative perspective on social movements' in D. McAdam et al. (eds).

McKibben, B, (2005), 'What the warming world needs now is art, sweet art', Grist [website], (21 April 2005), <http://www.grist.org/article/mckibben-imagine>, accessed 15 December 2009.

McCright, A.M and Dunlap, R.E. (2003), 'Defeating Kyoto: The conservative movement's impact on U.S. climate change policy' *Social Problems* 50:3, 348-373.

Meat Free Monday (MFM) (2010a), [website] <http://www.supportmfm.org/index.cfm>, accessed 13 May 2010.

Meat Free Monday (MFM) (2010b), 'Help the Planet' [website], <http://www.supportmfm.org/helptheplanet/>, accessed 13 May 2010.

Mellor, F. (2009), 'The politics of accuracy in judging global warming films', *Environmental Communication: A Journal of Nature and Culture* 3:2, 134-150.

Meyer, D. (2008), 'Building social movements', in S.C. Moser and L. Dilling (eds).

Meatless Monday (2010), 'The Movement goes global' [website], <http://www.meatlessmonday.com/the-movement-goes-global/>, accessed 1 May 2010.

Met Office (2009), 'Global temperature slowdown – not an end to climate change' [website], <http://www.metoffice.gov.uk/climatechange/policymakers/policy/slowdown.html>, accessed 20 December 2009.

Michaelis, L. (2006), 'Ethics of Consumption', in T. Jackson (ed).

Mies M. and Shiva, V. (1993), *Ecofeminism* (Halifax, Fernwood Publications: Canada).

Miller, D. (2008), *A Theory of Shopping* (Cambridge: Polity).

Monbiot, G. (2007), *Heat: How We Can Stop the Planet Burning* (London: Penguin).

Mormont, M. and Dasnoy, C. (1995), 'Source strategies and the mediatization of climate change', *Media, Culture and Society* 17, 49-64.

Moser, S.C. and Dilling, L. (2004), 'Making climate hot: Communicating the urgency and challenge of global climate change', *Environment* December 2004, 32-46

Moser, S.C. and Dilling, L. (eds) (2007), *Creating a Climate for Change: Communicating Climate Change and Facilitating Social Change* (Cambridge: Cambridge University Press).

Naranjo, L. (2002), 'Fragment of its former shelf', *Earth Observatory NASA* [website], (published online 28 May 2002), <http://earthobservatory.nasa.gov/Features/LarsenIceShelf/>, accessed 4 May 2009.

NASA (2009), 'Earth Observatory' [website] <http://earthobservatory.nasa.gov/>, accessed 4 May 2009.

Nature (2010), 'Editorial: Climate of suspicion', *Nature* 463 (21 January 2010), 269.

New Economics Foundation (Nef) (2009a), *The Consumption Explosion: The Third UK Interdependence Report* (London: New Economics Foundation).

New Economics Foundation (Nef) (2009b), *The Happy Index 2.0* (London: New Economics Foundation).

Newell, P. (2000), *Climate for Change: Non-state Actors and the Global Politics of the Greenhouse* (Cambridge: Cambridge University Press).

Nichols, B. (2001), *An Introduction to Documentary* (Bloomington: Indiana University Press).

Nicholson-Cole, S.A. (2005), 'Representing climate change futures: a critique on the use of images for visual communication', *Computers, Environment and Urban Systems* 29, 255-273.

Olausson, U. (2009), Global warming-global responsibility? Media frames of collective action and scientific certainty', *Public Understanding of Science* 18:4, 421-436.

O'Neill, S. and Nicholson-Cole, S. (2009), 'Fear won't do it: Promoting positive engagement with climate change through visual and iconic representations', *Science Communication* 30:3, 355-379.

Oreskes, N. (2004), 'The Scientific Consensus on Climate Change', *Science* 36, 3 December 2004, 1686.

Paul McCartney (2010), 'Charities: The Vegetarian Society' [website], <http://www.paulmccartney.com/charities.php#/1107/>, accessed 1 May 2010.

Pearce, F. (2007), *Earth Then and Now: Potent Visual Evidence of Our Changing World* (London: Mitchell Beazly).

Peta2 (2010), 'Meat's Not Green' [website], <http://www.peta2.com/meatsnotgreen/index.asp>, accessed 14 June 2010.

Pettenger, M.E. (ed) (2007), *The Social Construction of Climate Change: Power, Knowledge, Norms, Discourses* (Aldershot: Ashgate).

Pew Research Center (2009), 'Global Warming Seen as a Major Problem Around the World', Pew Global Attitudes Project, (2 December 2009), <http://pewresearch.org/pubs/1427/global-warming-major-problem-around-world-americans-less-concerned>, accessed 15 December 2009.

Pickerill, J. (2003), *CyberProtest: Environmental Activism Online* (Manchester: Manchester University Press).

Plumwood, V. (2002), *Environmental Culture: The Ecological Crisis of Reason* (London: Routledge).

Porter, H. (2009), 'The crushing of eco-protest brings shame on our police', *The Observer* [website], (19 April 2009), <http://www.guardian.co.uk/

commentisfree/2009/apr/19/police-powers-abuse-henry-porter>, accessed 20 April 2009.

Potter, S. (2008), *The Sting of Climate Change: Malaria and Dengue Fever in Maritime South East Asia and the Pacific Islands* (Sydney: Lowy Institute for International Policy).

Princen, T. and Finger, M. (1994), *Environmental NGOs in World Politics: Linking the Local and the Global* (London: Routledge).

Put People First (2009), [website], <http://www.putpeoplefirst.org.uk/>, accessed 10 December 2009.

Rees, W.E. (2008), 'Human nature, eco-prints and environmental justice' *Local Environment* 13:8, 685-701.

Robins, N. and Roberts, S. (2006), 'Making sense of sustainable consumption', in T. Jackson (ed).

Redclift, M. and Benton, T. (eds) (1994), *Social Theory and the Global Environment* (London: Routledge).

Risbey, J.S. (2007), 'The new climate discourse: Alarmist or alarming?', *Global Environmental Change* 18:1, 26-37.

Rootes, C. (ed) (1999), *Environmental Movements: Local, National and Global* (London: Frank Cass).

Rootes, C. (2005), 'A limited transnationalisation? The British environmental movement', in Donatella Della Porta and Sidney Tarrow (eds).

Rootes, C. (2006), 'Facing south? British environmental movement organisations and the challenge of globalisation', *Environmental Politics* 15:5, 768-786.

Rose, G. (2007), *Visual Methodologies: An Introduction to the Interpretation of Visual Materials* (London: Sage).

Royal Academy of Arts (2010), 'About the Royal Academy', [website], <http://www.royalacademy.org.uk/about/>, accessed 20 August 2010.

RSA Arts and Ecology (2010), [website], <http://www.artsandecology.org.uk/>, accessed 18 August 2010.

Sachs, W. (1999), *Planet Dialectics: Explorations in Environment and Development* (London: Zed Books).

Sample, I. (2007), 'Scientists offered cash to dispute climate study' *The Guardian*, 2 February 2007, <http://www.guardian.co.uk/environment/2007/feb/02/frontpagenews.climatechange>.

Sandler, R. and Pezzullo, P.C. (eds) (2007), *Environmental Justice and Environmentalism: The Social Justice Challenge to the Environmental Movement* (Cambridge, MA: The MIT Press).

Sandler, R. and Pezzullo, P.C. (2007), 'Introduction: Revisiting the environmental justice challenge to environmentalism' in R. Sandler and P.C. Pezzullo (eds).

Schiermeier, Q. (2010), 'The real holes in climate science', *Nature* 463 (21 January 2010), 284-287.

Schmidt, G. and Wolfe, J. (2009), *Climate Change: Picturing the Science* (New York: W.W. Norton and Company).

Science and Technology Committee, House of Commons (2010), *The Disclosure of Climate Data from the Climatic Research Unit at the University of East Anglia* (London: House of Commons, UK).

Seyfang, G. and Paavola, J. (2008), 'Inequality and sustainable consumption: bridging the gaps', *Local Environment* 13:8, 669-684.

Shackley, S. and Wynne, B. (1996), 'Representing uncertainty in global climate change science and policy: Boundary-ordering devices and authority', *Science, Technology and Human Values* 21:3, 275-302.

Shanahan, J. and Good, J. (2000), 'Heat and hot air: influence of local temperature on journalists' coverage of global warming', *Public Understanding of Science* 9:3, 285-295.

Shanahan, J. and McComas, K. (1999), *Nature Stories: Directions of the Environment and their Effects* (Cresskill, NJ: Hampton Press).

Shellenberger, M. and Nordhaus, T. (2004), 'The death of environmentalism: Global warming in a post-environmental world' (Oakland, CA: The Break Through Institute), <http://thebreakthrough.org/PDF/Death_of_Environmentalism.pdf>, accessed 16 October 2007.

Shellenberger, M. and Nordhaus, T. (2007), *Breakthrough: From the Death of Environmentalism to the Politics of Possibility* (New York: Houghton Mifflin).

Shellenberger, M. and Nordhaus, T. (2009), 'The emerging climate consensus: Global warming in a post-environmental world' (Oakland, CA: The Break Through Institute) <http://thebreakthrough.org/blog/PDF/EmergingClimateConsensus.pdf>.

Singer, P. (1995), *Animal Liberation* (London: Pimlico).

Slow Food (2010), [website], <http://www.slowfood.com/>, accessed 1 May 2010.

Slocum, R, (2004), 'Polar bears and energy-efficiency lightbulbs: strategies to bring climate change home', *Environment and Planning D: Society and Space* 22, 413-438.

Smith, H.A. (2007), 'Disrupting the global discourse of climate change: The case of indigenous voices', in M.E Pettenger (ed).

Smith, J. (ed.) (2000), *The Daily Globe: Environmental Change, the Public and the Media* (London: Earthscan).

Smith, N.W and Joffe, H. (2009), 'Climate change in the British press: the role of the visual', *Journal of Risk Research* 12:5, 647-663.

Smith, T.M. (1998), *The Myth of Green Marketing: Tending Our Goats at the Edge of Apocalypse* (Toronto: University of Toronto Press).

Soper, K. (1995), *What is Nature? Culture, Politics and the Non-human* (Oxford: Blackwell).

Soper, K. (2008), 'Alternative hedonism, cultural theory and the role of aesthetic revisioning', *Cultural Studies* 22:5, 567-587.

Spence, A. et al. (2010). *Public Perceptions of Climate Change and Energy Futures in Britain: Summary Findings of a Survey Conducted in January-March 2010. Technical Report* (Understanding Risk Working Paper 10-01). Cardiff: School of Psychology.

Stella McCartney (2010), [website], http://www.stellamccartney.com, accessed 16 June 2010.

Stop Climate Chaos (2010), [website], <http://www.stopclimatechaos.org/>, accessed 4 January 2010.

Sutton, P.W. (2000), *Explaining Environmentalism: In Search of a New Social Movement* (Aldershot: Ashgate).

Sweney, M. (2010), 'Ofcom to investigate government climate change TV campaign', *The Guardian*, (24 February 2010), <http://www.guardian.co.uk/media/2010/feb/24/ofcom-climate-change-bedtime-stories>, accessed 18 June 2010.

Szaz, A. (2007), *Shopping Our Way to Safety: How We Changed from Protecting the Environment to Protecting Ourselves* (Minneapolis: University of Minnesota Press).

The Guardian (2010), 'ABCs: National daily newspaper circulation December 2009', (15 January 2010), <http://www.guardian.co.uk/media/table/2010/jan/15/abcs-national-newspapers>, accessed 19 July 2010.

The Independent on Sunday (2007), '30 places you need to visit while you still can', *The Sunday Review, The Independent on Sunday*, 4 February 2007, 1.

Tickell, A. (2009), 'Introduction', in Julie's Bicycle (eds).

Trainor, S.F. et al (2007), 'Arctic impacts: Environmental injustice in Canada and the United States', *Local Environment* 12:6, 627-643.

Trumbo, C. (1996), 'Constructing climate change: Claims and frames in US news coverage of an environmental issue', *Public Understanding of Science* 5, 269-283.

Trumbo, J. (2000), 'Essay: Seeing science: Research opportunities in the visual communication of science', *Science Communication* 21:4, 379-391.

Ultimate Holding company (UHC) (2008), [website], <http://www.uhc.org.uk/>, accessed 18 January 2009.

UK Government (2009), 'Act on CO2', [website], <http://actonco2.direct.gov.uk/home.html>, accessed 17 August 2010.

UNFCCC (2009), 'Kyoto Protocol', [website], <http://unfccc.int/kyoto_protocol/items/2830.php>, accessed 12 June 2009.

UNFCCC (2010), 'The United Nations Climate Change Conference in Copenhagen, 7-19 December 2009', [website], <http://unfccc.int/meetings/cop_15/items/5257.php>, accessed 15 May 2010.

Ungar, S. (1992), 'The rise and (relative) decline of global warming as a social problem', *The Sociological Quarterly* 33:4, 483-501.

Ungar, S. (2000), 'Knowledge, ignorance and the popular culture: climate change versus the ozone hole', *Public Understanding of Science* 9, 297-312

Vidal, J. (2005), 'Revealed: how oil giant influenced Bush', *The Guardian*, (8 June 2005), <http://www.guardian.co.uk/news/2005/jun/08/usnews.climatechange>, accessed 7 May 2008.

Vidal, J. (2009), 'Copenhagen media coverage: A perfect storm', *The Guardian Media Section*, 7 December 2009, 3.

Wall, D. (1999), 'Mobilising earth first! In Britain', in C. Rootes (ed).

Weart, S. (2003), *The Discovery of Global Warming* (Cambridge, MA: Harvard University Press).

Weingart, P. et al. (2000), 'Risks of communication: discourses on climate change in science, politics and the mass media', *Public Understanding of Science* 9, 261-283.

Whitfield, P. (1997), *The Image of the World: Twenty Centuries of World Maps* (London: British Library).

Whitmarsh, L. (2009), 'What's in a name? Commonalities and differences in public understanding of "climate change" and "global warming"', *Public Understanding of Science* 18:4, 421-436.

Wilk, R. (2002), 'Consumption, human needs, and global environmental change', *Global Environmental Change* 12:1, 5-13.

Wilson, A. (1992). *The Culture of Nature: North American Landscape from Disney to Exxon Valdez* (Oxford: Blackwell).

Wolf, M.J.P. (1999), 'Subjunctive documentary: Computer imaging and simulation', in J.M. Gaines and M. Renov (eds), *Collecting Visible Evidence* (Minnesota: University of Minnesota Press).

World Development Report (2009), 'Public attitudes toward climate change: findings from a multi-country poll', (3 December 2009), <http://siteresources.worldbank.org/INTWDR2010/Resources/Background-report.pdf>, accessed 4 January 2010.

WorldWatch Institute (2010), *State of the World 2010, Transforming Cultures: From Consumersim to Sustainability* (New York: W.W. Norton).

WWF-UK (2008a), *Weathercocks and Signposts: The Environmental Movement at a Crossroads*, <http://www.wwf.org.uk/filelibrary/pdf/weathercocks_report2.pdf>, accessed 28 November 2008.

WWF-UK (2008b), 'Planet faces an ecological "credit crunch"' [website], <http://www.wwf.org.uk/news_feed.cfm?uNewsID=2294>, accessed 6 June 2008.

WWF-UK (2009), *Simple and Painless? The Limitations of Spillover in Environmental Campaigning*, <http://assets.wwf.org.uk/downloads/simple_painless_report.pdf>, accessed 14 March 2009.

WWF International (2008), [website], <http://www.oneplanetliving.org/index.html>, accessed 6 June 2008.

Index

References to illustrations are in **bold**

Adam, B. 3, 4, 6, 15, 26
  on atemporal time 25
  on clock time 26
  on role of science 19
Adams, C.J. 130, 136, 144
American Enterprise Institute (AEI) 18
Anderson, A. 28, 29, 94
Animal Aid (UK) 129, 134
animals, commodification 132, 144
Arctic/Antarctica, and climate change 51, 53
  see also Larsen B ice shelf
Atlantic Ocean, and climate change 49
Attenborough, David
  climate change documentaries 58-62, 63, 64, 65, 70
  Life on Earth 60

Barthes, R. 33, 43
BBC 32, 69-70
  Are We Changing Planet Earth? 58
  Can We Save Planet Earth? 58, 61
  The Heat is On 41
    credibility 40, 45
    visual imagery 39-40
  The Science of Climate 47
  The Truth About Climate Change 58
  Underwater Britain 53
Beck, U. 14, 15, 31, 105
  risk society thesis 20
Boykoff, M.T. 1, 14, 28, 38, 41, 47, 55, 101, 102, 105, 109
Brown, Gordon 106, 114, 118
Brundtland Report (1987) 127
Buckland, David, *Burning Ice* 151
Bunting, M. 153-4

CAFOD 79, 86-90
  action 87
  and climate change 87-88, 90
  FoEI, comparison 94
  hand imagery 88, **89**, 90
  values 86-87, 90, 99-100
Cameron, David 107
Camp for Climate Action 79, 81-82, 94-99
  communication practices 95-96
  expansion 98-9
  funding 95
  penknife design 96-97, **97**
    symbolism 98
  social values 98
  week-long camps 94-95
CAN *see* Climate Action Network
Cape Farewell (UK) 145
  aim 146
  Unfold exhibition 146, 152-153
carbon capture and storage 26, 64
  see also CO2
Caritas network 79, 86
Carvalho, A. 17, 18, 38, 47, 51, 102, 103, 105, 106, 120
Catholic Church, and climate change 86
climate
  measurement 16
  protest, and police actions 100
  socio-cultural dimensions 27
Climate Action Network (CAN) 77, 86, 91
climate change
  1870-1990 35
  actors, principal 23
  and Arctic/Antarctica 51, 53
  and the Atlantic Ocean 49
  and behaviour changes required 62-63
  and CAFOD 87-88, 90
  campaigning organizations 77-79, 81-82
    and social values 84

see also CAFOD; Camp for Climate Action; Friends of the Earth International
and the Catholic Church 86
and computer simulations 23
and consumption levels 123
and corporations 18
and the *Daily Mirror* 108-109
data, observed/simulated 56, 68
and deforestation 129
and flooding 53-54, 61, 109
FoEI, justice discourse 91-92
and fossil fuels 47, 49
and human agency 56, 57, 60, 61, 67
as humanitarian issue 8
images
culturally meaningful 148-149
dominant 61, 69
melting glaciers 1, 24, 32, 59, 63, 136, 145, 148
polar bears 53, 55, 61, 145, 153
role of 31-32
typology 32, 54, 58, 66, 69
individualist discourse 64
invisibility of 4, 5, 21, 31, 71
IPCC role 105
and mainstream media 101
mainstreaming of 1-2
manifestations 38
materialization 25
and meat consumption 133, 134, 140
media coverage 28-29
mediation of 2
and NGOs 77-78
role 38-9
public perceptions of 24
reversibility, assumed 26
risks 20
scientific consensus on 14
undermining 110-111
scientific uncertainty 19, 40-41, 105
social movements 83, 99-100
and soybean production 138
stability, assumed 16
timescales 62, 68-69
and UK floods 53
use of term 41
visible signs 1, 24, 54-55
and the visual arts 145-155
visual communication of 31, 44, 154-155
and weather events 53, 67-68, 109
see also global warming
Climate Change Bill (UK) 93
Climate Justice Action 119
Climate Justice Now! 77, 82, 91
climate modelling 33-34, 45, 46, 49
CO2 effects 36
Greenpeace 50
uncertainties 34, 108
climate sceptics 38, 106-107
climate science 34, 57, 66, 102
Daily Mail, scepticism about 112
presumed rationality 106
uncertainties 108
'Climategate' controversy 1, 13-14, 30, 41, 93, 110
at COP15 102, 105, 120
and *Guardian* 106
and *Independent* 108
Climatic Research Unit (UEA) 1, 13
CO2
campaign, UK Government 157
effects, climate modelling 36
emissions
and increased temperatures 38
time lapse 57
production 1, 4
Collins, Sam, *Sometimes the Journey is Better than the Destination* 152
Compassion in World Farming, campaign against meat consumption 129
computer simulations 34, 44, 62, 68, 69
consumerism
consumption, distinction 125-126
ideology 126
consumption
alternative forms 126-127
consumerism, distinction 125-126
global inequities 128
and happiness 128
levels, and climate change 123
sustainable
ethics of 124, 125, 126, 142-144
meaning 127-128
WWF on 128

Copenhagen Accord 104
Copenhagen Conference (COP15) (2009) 2, 77, 83, 92
  agenda 104-105
  civil society groups 119-120
  'Climategate' controversy 102, 105, 120
  context 103-104
  newspaper reporting 103, 104-105, 113-118, 121-122
corporations, and climate change 18
Cosgrove, D. 23, 46, 47, 90
Cottle, S. 6, 9, 28, 82, 95, 118, 119, 120, 148
Cox, P. 60-61, 62, 68
Cox, R. 3

Daily Mail, climate science, scepticism about 112
Daily Mirror, and climate change 108-109
dairy products, consumption *see* meat and dairy consumption
Dasnoy, C. 23, 32, 39, 43, 77
Davis, David 107
deforestation, and climate change 129
DeLuca, K. 5, 23
Diani, M. 80, 81, 87, 94
Disaster and Emergencies Committee (DEC) 86
Donati, P.R. 80, 81, 87, 94
Doyle, Julie *see* Harradine, David

Earth Observatory, NASA 22
Earth Summit (1992) 147
EarthFirst! 81
environment
  campaign groups 23
  conceptualisation 3-4, 21-27, 78
ExxonMobil 18, 65

factory farming, case against 138
flooding, and climate change 53-54, 61, 109
Food Chain Campaign (FoE) 138-140, 141
  aims 130
  government regulation, focus 140, 143
  images 139, **139**
  video 138-139
fossil fuels, and climate change 47, 49
Friends of the Earth International (FoEI) 23, 79, 91-94
  CAFOD, comparison 94
  climate change
    images 92
    justice discourse 91-92
  Get Serious About CO2 campaign 93, 99
  humorous/serious strategy 93
  membership 91
  MOVIAC 93
  risk, appeal to 93
  Safe Climate campaign 93
  social networking use 93-94
  structure 91
Friends of the Earth (UK), Food Chain Campaign 130, 138-140

Get Serious About CO2, FoEI 93, 99
glaciers, melting, climate change imagery 1, 24, 32, 59, 63, 136, 145, 148
Global Climate Coalition (GCC) 14, 65-66
global south 82fn3
global warming 16-17, 18
  invisibility 28
  and meat consumption 135
  and scientific evidence 40
  use of term 41
Gore, Al, *An Inconvenient Truth* 1, 55, 58
Gormley, Anthony, *Amazonian Field* 147
Green Alliance 84
greenhouse gas emissions 2, 57
  and livestock farming 124, 128-129
  and meat/dairy consumption 128, 129, 141
  responsibility for 123
  stabilisation 63
Greenpeace 1, 23, 32
  climate change
    communication 41-43, 48
    images 68, 69, 70
  climate modelling 50
  publications
    Emerging Impacts of Climate Change? 39, 41, 44
      images **42**, **44**
    How to Save the Climate 67
      front cover **65**

Nature's Bottom Line 51-52
front cover **52**
Putting the Lid on Fossil Fuels 47-48, 49, 52
front cover **48**
Guardian, and 'Climategate' controversy 106

Hajer, M. 2, 127
Hansen, A. 4, 5, 23, 29, 46, 53, 94, 102
Hansen, J. 106
Harradine, David, and Julie Doyle 154
Here Today: Moving Images of Climate Change 155
Hatoum, Mona, *Hot Spot* 147-148
Hobson, K. 63, 64, 127, 128, 133, 142
Hulme, M. 2, 15, 16, 20, 26, 27, 84
human agency, and climate change 56, 57, 60, 61, 67
Hurricane Katrina 60, 61

identity campaigning 85-86
Independent
and climate sceptics 107
and 'Climategate' controversy 108
Indigenous Environmental Network 119
International Geophysical Year (1957-8) 37
Internet, and environmental activism 82-83
IPCC
climate change, role 105
criticism of 19
membership 17-18
Reports 13, 18, 24, 26, 32
1990 33, 37, 38
1995 46-7
2001 51, 55
2007 53, 55
maps 37
synthesis 57

Jasanoff, S. 36-37, 46, 69, 74

Klimaforum09, declaration 82
Kyoto Protocol (1997) 1, 5, 38, 47, 103, 104
entry into force 51
targets 51

Larsen B ice shelf 51-52, **52**, 73-74, **73**
Latour, B. 4, 15, 30, 78
Leiserowitz, A. 1, 24, 30, 31
Lester, L. 5, 24, 31, 82, 93
livestock farming, and greenhouse gas emissions 124, 128-129

McCartney family, professional credibility 131-132, 134
McCartney, Linda 131
meat free recipes 132
McCartney, Mary 131
McCartney, Paul 131
McCartney, Stella 131
McKibben, Bill 145
Macnaghten, P. 3, 5, 22, 145
Manzo, K. 32, 43, 45, 54, 58, 66, 69, 87, 88, 89, 90, 92, 99
maps, illusion of mastery 36
meat
attitudes to 130
consumption
and climate change 133, 134, 140
Compassion in World Farming campaign 129
and global warming 135
and dairy consumption
campaigns against 128, 130, 141-142
and greenhouse gas emissions 128, 129, 141
symbolic meaning 130
Meatless/Meat Free Monday campaign 130-134
ethical appeal 140-141
meat free recipes 132
website 131-133
Meat's Not Green campaign (Peta2) 135-137
animal images 136
individualist discourse 137, 140
video 136
website 135-136
Miliband, Ed 106
Moffatt, Tracey, *Doomed* 149
Mormont, M. 23, 32, 39, 43, 77
Movement of Victims Affected by Climate Change (MOVIAC) 93

NASA
earth images 37, 74
Earth Observatory 22
National Center for Atmospheric Research (NCAR) 39
nature
aestheticising 22
commodification of 21, 125
representations of 22
and vision 21-24
Newell, P. 14, 19, 77, 105
newspaper reporting
COP15: 103, 104-105, 113-118, 121-122
undermining of scientific authority 112-113
NGOs
and climate change 77-8
environmental 81
role 38-39
websites 82
Nicholson-Cole, S. 24, 29, 31, 55, 70, 127, 145
nuclear power
as carbon free 64
promotion of 111

Obama, Barack 102
O'Neill, S. 24, 29, 31, 33, 55, 70, 101, 102, 127, 145
ozone hole 29

People for the Ethical Treatment of Animals (PETA) 120, 134
Peta2, Meat's Not Green campaign 135-137
photography, Greenpeace use of 45
Picala, Steve 63
PLATFORM (UK) 145
Plumwood, V. 3, 4, 19, 20, 21, 30, 85, 105, 124, 130, 137, 143, 144
polar bears, climate change imagery 53, 55, 61, 145, 153
police actions, and climate protest 100
Prescott, John 54

Reay, Dave 63
RETHINK: *Contemporary Art and Climate Change*, exhibition 145
risk society thesis, Beck 20
Royal Academy of Arts, *Earth: Art of a changing world* 145, 146, 151

Sachs, W. 34, 36, 37, 46, 69
Schneider, Stephen 39, 40
science
duality of 20
politicization 106
role, Adam on 19
scientific
authority, undermining 110-111, 112-113
evidence, and global warming 40
knowledge, as truth 14-15
rationality 45
uncertainty, and climate change 19, 40-41, 105
Shackley, S. 14, 15, 19, 20, 34, 38, 40, 105, 122
social movements
climate change 99-100
definition 80
framing processes 80
mobilising structures 80
political opportunities 80
purpose 79
theories 79
social networking, use by FoEI 93-94
social values
campaigning organizations 84
WWF on 84-85
Soper, Kate 127
soybean production, and climate change 138
Stern Review (2006) 55
neoliberal agenda 67
Stop Climate Chaos Coalition (UK) 86
Szaz, Andrew 126

Thatcher, Margaret 17
theory, and action 6
Tickell, Alison 153
time
atemporal, Adam on 25
clock, versus environmental time 25, 26
Times, The, scientific authority, undermining 110-111

Twomey, Claire, *Specimen* 150-151

UK
floods
Boscastle 61, 109
and climate change 53
Government campaign, CO2 157
Youth Climate Coalition 119
UN
Earth Summit (1982) 101
Framework Convention on Climate Change (1992) 51
UNEP 13
Urry, J. 4, 5, 22, 145

veganism 144
Vegetarian Society 131, 132
Vidal, John 101
vision, and nature 21-24
visual arts, and climate change 145-155

Weart, R. 16, 17, 18
weather events, and climate change 53, 67-68, 109
Whitmarsh, L. 1, 41
WMO 13, 16, 27
Women's Environmental Network 81
World Climate Conference (1979) 16, 17
WWF 23, 63, 83
on consumption 128
on social values 84-85
Weathercocks and Signposts 84
Wynne, B. 14, 15, 19, 20, 34, 38, 40, 105, 122